IMPROVING ENGINEERING DESIGN

Designing for Competitive Advantage

Committee on Engineering Design Theory and Methodology
Manufacturing Studies Board
Commission on Engineering and Technical Systems
National Research Council

NATIONAL ACADEMY PRESS
Washington, D.C.   1991

This study was supported by Contract No. DMC-8817926 between the National Science Foundation and the National Academy of Sciences.

Library of Congress Card Catalog No. 91-60349   First Printing, February 1991
ISBN 0-309-04478-2   Second Printing, January 1992
                                                         Third Printing, May 1992

A limited number of copies are available from:

Manufacturing Studies Board
National Research Council
2101 Constitution Avenue, Room HA270
Washington, D.C. 20418

Additional copies are available for sale from:

National Academy Press
2101 Constitution Avenue
Washington, D.C. 20418

S-323

Printed in the United States of America

Preface

Effective design and manufacturing, both necessary to produce high-quality products, are closely related. However, effective design is a prerequisite for effective manufacturing; quality cannot be manufactured or tested into a product, it must be designed in. The United States needs to sharpen its understanding of engineering design theory if it is to realize the competitive advantages of superior engineering design. Significant improvement of design practice requires increased knowledge of the fundamentals of design and increased readiness of firms to adopt new methods. Developing and teaching a coherent body of engineering design principles in this area could help accelerate the changes necessary to maintain the competitiveness of future U.S. manufacturing.

This report presents the findings and recommendations of the Committee on Engineering Design Theory and Methodology, formed by the Manufacturing Studies Board of the National Research Council at the request of the National Science Foundation. The scope of the committee's efforts was to:

Determine the importance of engineering design to U.S. industry's competitiveness in world markets;

Articulate the means by which the practice of engineering design in the United States can be improved;

Propose actions to improve undergraduate and graduate education in engineering design;

Propose a national effort to improve the practice of engineering design through research and development; and

Recommend to government, industry, and academe mechanisms for improving engineering design practice, education, and research.

The committee, consisting of 16 experts in the primary fields of engineering design—education, practice, management, and research—worked in part as three subcommittees to explore the status of engineering design practice, education, and research in the United States. The committee has based this report on its discussions and analysis of the current environment for engineering design; as such, it reflects the consensus of the committee on the implications of engineering design in the United States.

This report was enabled by many people directly and indirectly at work on engineering design. The study was conceived and planned by John Dixon and Michael Wozny of the NSF and George Kuper and Kerstin Pollack of the NRC. Site visits to the following companies contributed to a greater understanding of issues in the practice of design: American Precision Industries, AT&T Bell Laboratories, Cooper Industries, Ford Motor Company, General Electric Company, Hewlett-Packard, and Polaroid. Many engineering deans and design faculty contributed by describing their current engineering design research, industrial applications thereof, predicted developments, and potential barriers. The contributions to the committee's deliberations of Karl Ulrich, assistant professor, Sloan School of Management, Massachusetts Institute of Technology, deserve special attention. Main staff support was ably provided by Paul Shawcross, with Janice Greene and Kerstin Pollack providing key help and Lucy Fusco playing a strong supporting role. Theodore Jones assembled the report, and Kenneth Reese edited it.

Charles W. Hoover and J. B. Jones
Co-Chairmen, Committee on Engineering
Design Theory and Methodology

Contents

IMPROVING ENGINEERING DESIGN

Executive Summary

Engineering design is a crucial component of the industrial product realization process. It is estimated that 70 percent or more of the life cycle cost of a product is determined during design. Effective engineering design, as some foreign firms especially have demonstrated, can improve quality, reduce costs, and speed time to market, thereby better matching products to customer needs. Effective design is also a prerequisite for effective manufacturing. Improving the practice of engineering design in U.S. firms is thus essential to industrial excellence and national competitiveness.

Unfortunately, the overall quality of engineering design in the United States is poor. The best engineering design practices are not widely used in U.S. industry, and the key role of engineering designers in the product realization process is often not well understood by management. Partnership and interaction among the three players involved in this endeavor—industries, universities, and government—have diminished to the point that none serves the needs of the others. Engineering curricula focus on a few conventional design procedures rather than on the entire product delivery process, and industry's efforts to teach engineering design tend to be fragmented. A revitalization of university research and teaching in engineering design has begun, but is not well correlated with the realities or scope of design practice, and research results are not effectively disseminated to industrial firms. Finally, the U.S. government has not recognized the enhancement of engineering design capabilities to be of national importance.

This state of affairs virtually guarantees the continued decline of U.S. competitiveness. To reverse this trend will require a complete rejuvenation of engineering design practice, education, and research, involving intense cooperation among industrial firms, universities, and government.

1

DESIGNING FOR COMPETITIVE ADVANTAGE

To use design effectively as a tool for turning business strategy into effective products, a firm must (1) commit to continuous improvement both of products and of design and production processes, (2) establish a corporate product realization process (PRP) supported by top management, (3) develop and/or adopt and integrate advanced design practices into the PRP, and (4) create a supportive design environment.

Converting to operation under the discipline of a PRP is not easy. Often, complete reorganization from top to bottom and a dramatic change in the way of doing business are required. An effective PRP generally incorporates the following steps: define customer needs and product performance requirements; plan for product evolution beyond the current design; plan concurrently for design and manufacturing; design the product and its manufacturing processes with full consideration of the entire product life cycle, including distribution, support, maintenance, recycling, and disposal; and produce the product and monitor product and processes.

The PRP is a firm's strategy for product excellence and continuous improvement; design practices are its tactics. Because not all practices are applicable to or useful in the design of a given product, each company must carefully identify a set appropriate to its uses and incorporate them into its PRP. Practices (such as Taguchi methods) and tools (such as CAD and CAE) must be fully integrated into the PRP if they are to have more than minimal effect. Companies must also develop means of assimilating new practices as they are developed by researchers and others because currently effective practices are being improved and even superseded.

Design is a creative activity that depends on human capabilities that are difficult to measure, predict, and direct. An understanding of the design task and the characteristics and needs of people who design effectively is essential to the creation of a stimulating and nurturing design environment.

IMPROVING ENGINEERING DESIGN EDUCATION

Undergraduate and graduate engineering education is the foundation for successful practice, effective teaching, and relevant research in engineering design. The current state of that foundation is attested to by employers who find recent engineering graduates to be weak in design. Reasons for the inadequacy of undergraduate engineering design education include: weak requirements for design content in engineering curricula (many institutions do not meet even existing accreditation criteria); lack of truly interdisciplinary teams in design courses; and fragmented, discipline-specific, and uncoordinated teaching. Of the curricula that have strong design components, few consider state-of-the-art design methodologies.

There are simply too few strong graduate programs focusing on modern design methodologies and research to produce the qualified graduates needed by both industry and academe. Limited funding for design research impairs the quality of graduate programs in design and reduces the number of graduate students in the field that can be supported. Even the stronger programs rarely involve industry experience that would elucidate the realities of engineering design practice.

Significant improvement in engineering design is unlikely without strong, knowledgeable, enthusiastic faculty who interact with a broad base of colleagues in industry as well as academe. However, few faculty today are trained to teach design or are cognizant of its importance. Most have no significant industrial design experience, possess little understanding of manufacturing, and have only limited contacts with industry. Relevant textbooks are lacking, and many faculty are unfamiliar with the instructional techniques that best support design education. Faculty who would consider design as a career focus face a significant time commitment and institutional obstacles.

The initiative for immediate improvement of design education and for laying the groundwork for its longer-term sustained improvement lies clearly with educational institutions. Faculty and administrators, who sometimes disclaim responsibility for the problem and blame instead the "system," must take the lead if it is to change. To improve the teaching of engineering design in universities will require: recognition of the deficiencies in design education; strong high level leadership in establishing goals for improving design education; development of metrics to measure progress toward these goals; creation of designated change agents to plan and implement improvements; and extensive training programs for both new and experienced design teachers.

Actions must also be taken to facilitate the teaching of design and to increase university-industry cooperation in design education. A national clearinghouse for design instructional materials could make the task of teaching design easier for many faculty. Industrial firms could help improve engineering design by encouraging faculty to work in industry, aiding universities in setting goals and planning curricula, and supporting research in engineering design.

A NATIONAL AGENDA FOR ENGINEERING DESIGN RESEARCH

Research is a central ingredient in repairing the national infrastructure in engineering design. It will contribute new knowledge, new ideas, and new people to industry and education and stimulate the creation of new business enterprises. Over time, a well-conceived, sustained program of engineering design research will gradually reduce U.S. companies' reliance on ad hoc design methods and improve their ability to produce higher-quality, lower-cost products and reduce lead time to market for new or modified products.

Ten topics in three broad areas—developing scientific foundations for design models and methods, creating and improving design support tools, and relating design to the business enterprise—were deemed crucial to reforming the practice and teaching of engineering design. Collectively, they comprise a national research agenda that will serve to guide the National Science Foundation, other government agencies, private foundations, industrial firms, and individual researchers in the assignment of research priorities and selection of projects.

The proposed research is essential to the revitalization of the engineering design infrastructure in the United States and hence to U.S. competitiveness. Significant and useful intermediate (i.e., four- to five-year) results should be achievable for most topics. It is extremely important that this research, whether applied or basic, be of the highest quality and be conducted with frequent and close interaction between researchers and industry design engineers, and that results be disseminated to industry as well as to academe.

Results of university research in engineering design can find their way into industrial practice by a number of routes. However, even well developed research results cannot simply be "given" to industry; new methods must be refined and packaged as products, a task that cannot readily be performed by most universities or by most companies that might take advantage of the results. The creation of a National Consortium for Engineering Design (NCED) to perform this technology transfer role should be considered.

RECOMMENDATIONS

Industrial design practice, engineering education, and design research all can be improved. Many of the report's recommendations require only initiative by the actors and little investment. Companies must reorganize their product realization processes and at least adopt existing best design practices. They must also communicate better with universities in order to secure new design methods and well-prepared graduates. Universities, in turn, must make a high-level commitment to improve engineering design education and research and better relate them to the needs of industry. The government must make engineering design a national priority and encourage research by increasing funding and assisting in the establishment of clearinghouses for design information and teaching materials. Specific actions are recommended in the report.

1

Introduction

It is now widely believed that U.S. industry's extended period of world dominance in product design, manufacturing innovation, process engineering, productivity, and market share has ended.[1] The once globally dominant U.S. automobile and steel industries have lost market share at home and abroad, and U.S. products have all but disappeared from the consumer electronics market. There is consensus that U.S. industry as a whole is not as productive as it might be, and that its rate of productivity increase is lower than that of industries in many other nations.[2] This loss of competitiveness with foreign firms has been keenly felt in some areas in job losses and plant closings. Profitability continues to decrease in many key industries, threatening further loss of market share and jobs. U.S. citizens, from the individual consumer to the senior corporate executive, daily observe evidence of the decline of the nation's "industrial might."[3] Figures 1 and 2 illustrate the declining performance of some important U.S. industries.

The decline of U.S. international competitiveness has been ascribed to many factors, among them national fiscal and trade policies, exchange rates, national "culture," deficiencies in manufacturing, industrial management and accounting practices, unfair foreign trade practices, and methods of providing capital. A crucial factor that is not often recognized is the quality of engineering design in U.S. industry. Engineering design is the key technical ingredient in the product realization process (PRP),[4] the means by which new products are conceived, developed, and brought to market. (Various other names, including *concurrent engineering*, are in use for the product realization process or for major parts of it.) The ability to develop new products of high quality and low cost that meet customer needs is essential to increasing profitability and national competitiveness. The link between quality and profitability has been convincingly demonstrated by studies using the PIMS[5]

5

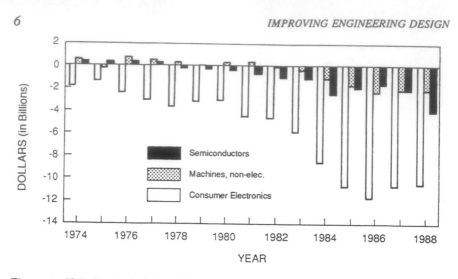

Figure 1: U.S. Trade Deficit in Three Key Industries

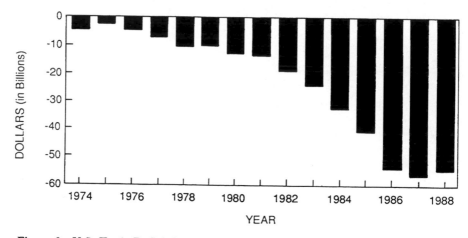

Figure 2: U.S. Trade Deficit in the Auto Industry

data base. Figure 3 summarizes the results of a study done using the PIMS data base that shows the effects of quality and market share on profitability for a large group of U.S. industries, predominantly manufacturers.[6]

THE CENTRAL ROLE OF ENGINEERING DESIGN

High-quality products satisfy customer needs for reliability, serviceability, and acceptable life cycle cost, as well as for functionality and aesthetics. Competitiveness demands high-quality products, which require high quality

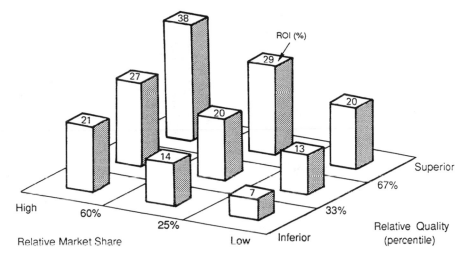

Figure 3: Return on Investment as a Function of Quality and Market Share

in their components and in the systems and processes used in their production. Effective design and manufacturing, both necessary to produce high-quality products, are closely interrelated, but effective design is a prerequisite for effective manufacturing; quality cannot be manufactured or tested into a product, it must be designed into it.[7] Figure 4, derived from studies done at Westinghouse and General Motors, suggests that a major fraction of the total life cycle cost for a product is committed in the early stages of design.[8]

As products become more complex, containing more and more parts, manufacturing yield falls dramatically unless design efforts can create parts and manufacturing operations of extremely high quality. This sensitivity of final product quality to component quality as complexity increases may be readily demonstrated. Assume that a final product requires n components and operations, each with a probability of being acceptable, P_j. Then the probability of the final product being acceptable, P, is

$$P = \prod_{j=1}^{n} P_j \qquad (1)$$

If $P_j = p$ is the same for all n components and operations, then equation 1 simplifies to

$$P = (p)^n \qquad (2)$$

Figure 5 is a parametric plot of equation 2 which shows that very high quality in all components and assembly operations is required to get accept-

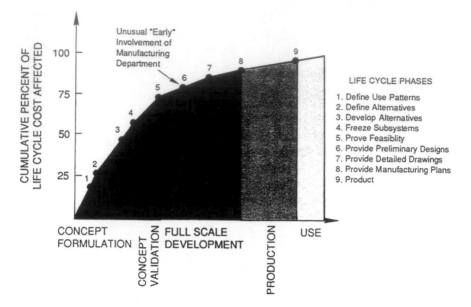

Figure 4: Life Cycle Cost Commitment

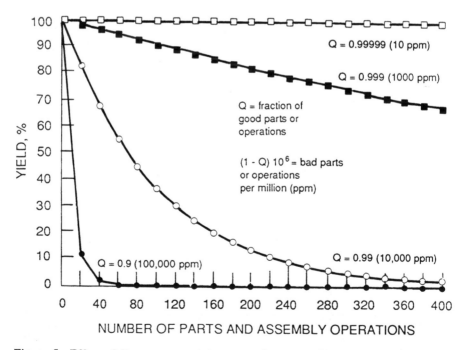

Figure 5: Effect of Component and Assembly Quality on Yield

able yields for products with even a few hundred parts or assembly operations. Note that a component quality of 10 ppm (defective parts per million parts) is required to get yield in the 99 percent range for a system composed of 400 parts.

U.S. performance in engineering design can be compared to that of other nations on the basis of the speed and cost with which new product concepts and product improvements are brought to market and customer perceptions of the quality and performance of those products. The greater time from concept to delivery for U.S. than for Japanese products is illustrated by Figure 6.[9]

Manufacturing performance, including adherence to design specifications, flexibility, and efficiency, is also involved, but effective design is at the heart of the concept of continuous accumulated improvement—the drive to make a product better year after year.

When measurements are made, it becomes clear that U.S. industry's loss of market share in many industries results from poor performance in the very areas in which successful foreign companies, particularly some Japanese companies, usually excel.[10] Loss of market share resulting from poor design is likely to spread as foreign competition expands into other industries— aerospace, large appliance, and cosmetics industries being likely near-term targets.

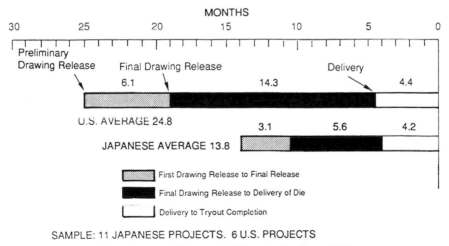

SAMPLE: 11 JAPANESE PROJECTS. 6 U.S. PROJECTS

The totals may not add because of missing values in subsequent data

Figure 6: Lead Time For a Major Body Die (Months)

THE NATURE OF ENGINEERING DESIGN

As the key technical ingredient in the product realization process, engineering design bears responsibility for determining in detail how products will be made to meet performance and quality objectives for the customer at a cost that permits a competitive price. It thus plays a key role in the ability of businesses to excel.

Engineering design has both technological and social components. The technological component includes knowledge about engineering science, design methods, engineering models, materials, manufacturing, and computers. The social component includes corporate organization and culture, team design methods, the nature of the design task and of the designer, customer attributes, and employee involvement.

An ever-evolving problem-solving activity, engineering design encompasses many different and increasingly advanced practices, including methods for converting performance requirements into product features, computer-integrated manufacturing, cross-functional teams, statistical methods, competitive benchmarking of products, computerized design techniques, and new materials and manufacturing processes. These and other methods used by the most competitive companies worldwide do not exist or operate independently, but rather are integrated into a unified process.

The committee considered a broad range of engineering design activities, including practices, processes, principles, methodologies, and techniques employed in companies large and small. Although the committee did not focus on Very Large Scale Integration design or software design because these are narrower domains, significant successes in these areas are ascribed to the close coupling of product and process design and thus provide lessons for all areas of design.

Findings—The Current State of Engineering Design in the United States

Several committee members had past or ongoing professional experience in key roles in improving design practice in their respective companies. These committee members had benchmarked their firms against leading competitors and often found their firms wanting. Significant benchmarks often considered included factors such as cycle time, the number of iterations of the design cycle, and the number and administration of design changes. Starting with this background, a panel of these members developed a set of questions which they posed to a number of leading-edge companies. They found that these U.S. firms believe significant efforts will be needed to attain the advantages that already accrue to their most effective foreign competitors, who successfully apply advanced design practices. In addition, attaining a similar level of competence involves a moving target. The companies visited were large and well supported. The committee did not explicitly study small and medium-sized firms, whose situation is starkly documented in various reports that show that their level of adoption of even computer-aided design (CAD) substantially lags that of foreign firms.

The status of research in design theory was assessed by a different approach. A panel of experts in the field drafted, refined, and ranked a set of topics covering the various areas of investigation. They estimated the minimum support necessary to get "above-threshold" progress in these areas and compared these desired levels to existing levels of support. They concluded that current levels of support are far less than needed to advance the field.

Addressing design education, the committee once again drew upon the extensive experience of its members, but also visited and posed questions to industrial firms on the adequacy of the education of newly hired engineers.

The following general statements are offered with no intent to cover all cases; subsequent sections of this report identify wide variations in industrial and educational practice throughout the nation. Nevertheless, the committee's findings support the following statements on the current state of engineering design in this country.

1. **The best engineering design practices are not widely used in U.S. industry.**[11] Many U.S. companies limited by existing practices are unwilling to try new ones, often because of management rather than technical barriers. Those U.S. companies that do try to identify and absorb current best practices are still often outstripped by their best foreign competitors, which continue to evolve new and still better practices. A higher rate of new product introduction in these foreign firms results in more rapid learning, which translates into more rapid improvement of design and manufacturing processes.[12] Improvement migrates slowly in the United States because the process of sharing and disseminating design knowledge among companies remains dependent on informal networking of individuals.

2. **The key role of designers in the PRP is often not well understood by management.** Most designers take on, often by default and without portfolio, an enormous range of new activities in support of the PRP, and management often does not recognize the importance of these nontraditional design activities. Motivation and support of designers is complicated because there is no way to use data from traditional cost accounting systems to evaluate the contribution of design to profit or to compare the effectiveness of different designs. In recognition of this problem, proposals for different cost accounting systems have recently been published.[13]

3. **Some U.S. firms use design effectively, but they have had to change their goals and culture to do so.** To move from stable, high volume, slowly changing production to continuous improvement requires profound cultural

change; firms that have made this shift have adopted an all-enterprise approach, employing dedicated agents to catalyze and support change. These firms use a product realization process as the vehicle for involving people at all levels and in all functions in defining, designing, and producing the product and moving it to market. They choose design practices to support the PRP and design the product, and they set metrics to guide the process.

4. **Partnerships and interactions among industry, research, and education are so limited that the relevant needs of each are poorly served by the others.** With few exceptions, engineering design education and research is divorced from industry needs. For its part, industry does not articulate its requirements, support changes in the design component of curricula, or view education as an incubator of design talent. University design research efforts are often isolated from industry, and industry rarely uses the results of university research. Although some companies have fared well despite this environment, most (particularly medium-sized and small companies) suffer the consequences of outdated methods and poorly prepared new engineers in product quality, market share, competitiveness, and international trade.

5. **Current engineering curricula do not focus on the entire product realization process.** Most curricula emphasize a few steps of conventional, essentially technical, design procedures. Curricula as a whole lack the essential interdisciplinary character of modern design practice and do not teach the best practices currently in use in the most competitive companies. The result is engineering graduates who are poorly equipped to utilize their scientific, mathematical, and analytical knowledge in the design of high-quality components, processes, and systems. Few have experienced design as part of a team, even fewer understand the multiple goals that motivate design, and most lack sufficient understanding of statistics, materials, manufacturing processes, cost accounting, and product life cycle considerations. Industrial training courses try to fill these gaps at considerable cost and with varying degrees of success.

6. **Industry's internal efforts to teach engineering design, intended to compensate to some degree for these shortcomings, are too fragmented and not institutionalized as natural components of the way business is performed.** These efforts, affordable only by the largest companies, are not based on the fundamental understanding of design processes that could be provided by design research. Yet most engineers, including new employees, currently learn modern design techniques from industrial training courses.

7. **Although universities nominally bear responsibility for producing both practices and practitioners, they do not fulfill this role in engineering design in the United States.** The breakdown extends beyond curricula. Universities do not, in general, value engineering design as an intellectual activity, either in research or in teaching. Lack of instructional materials and experienced faculty and the need for time-consuming interaction with

students make courses in design difficult to teach. Many who do teach design have little experience and are unaware of the most recent design techniques. The few efforts to revitalize university research and teaching in engineering design are fragmented, insufficiently funded, and not well enough coupled to the needs of industry to produce either well-prepared new engineers or useful research results.

8. **A revitalization of university research in engineering design has begun.** Unfortunately, it is not well correlated with the realities of the full scope of design for competitive products, and results are not well disseminated to industry. The National Science Foundation's (NSF's) program in engineering design theory and methodology is funded at too low a level and not yet recognized by the research community as a stable source of research leadership and support. NSF's Engineering Research Centers, some of which have design-oriented research thrusts, are a step in the right direction, but again, funding for design efforts is inadequate.

9. **The U.S. government has not recognized the development of superior engineering design as a national priority.** Though engineering design is a primary determinant of competitiveness over the entire spectrum of manufacturing industries, it has not received the level of support that has been accorded specific product areas such as semiconductors and superconductors.

This state of affairs virtually guarantees the continued decline of U.S. competitiveness over the long term. A complete rejuvenation of engineering design practice, education, and research—aimed at future needs rather than just at "catching up" to competitors' current standards—is fundamental to gaining and maintaining U.S. industrial competitiveness. An objective of this magnitude requires intense cooperation among industries, universities, and the government.

In the United States, federal and state government policies have not traditionally been directed toward helping private enterprises enhance their competitiveness through adoption of advanced technologies, in part because technology-based industries have in the past faced little serious competition from foreign firms. Now nearly all foreign competition in high-value-added products is strengthened to some extent by various foreign government measures to increase the technological strength of key industries. Consequently, traditional government policies warrant intense restudy and, in all likelihood, revision.[14]

THE CONSEQUENCES OF BETTER DESIGN PRACTICE, EDUCATION, AND RESEARCH

Improving engineering design practice in U.S. industry will result in shorter development time, lower cost, and better match of products to customer wants. The fastest way to realize these benefits is for the vast majority of U.S. companies to learn to use the advanced design practices that have already been implemented by leading-edge companies in the United States and abroad. It has taken these pioneering companies five to eight years to change their practices, yet many are willing to share their lessons, enabling other companies to learn and implement advanced design practices in a much shorter time.

On a slightly longer time scale, better engineering design education will improve the practice of engineering in the United States. If the committee's recommendations are followed, in a few years universities will begin to graduate students whose knowledge of engineering design, contact with industry during their schooling, and awareness of good design practices will better attune them to the needs of industry and the realities of engineering design and dispose them to continuing education throughout their careers. These graduates will augment and eventually replace a generation of designers who received limited coherent engineering design education. Students who emerge from graduate engineering design programs familiar with current advances in theoretical foundations of design and forefront methodologies will not only contribute to engineering practice, but also be prepared to create new design tools, teach design to next generation students, and conduct research in design.

The benefits of expanded design research will take longer to accrue—even with improved dissemination of research results to U.S. industry and greater eagerness on the part of industrial firms to use the results—but may have the greatest impact on productivity. Indeed, given the best result, it could provide the means for leaping ahead of the competition. Research will provide new design methods and principles to support more rapid development of further improved design practices. It will provide tools for faster and more complete learning of design methods by both practicing engineers and students, multiplying both the quantity and quality of design engineers. Research results will be further developed into computer programs, data bases, visualization devices and techniques, methods of predicting behavior and cost early in the design process, and other valuable, but today unforeseeable, mechanisms.

It is crucial that improvements be made in each of the three areas of design—practice, education, and research. Halfway measures will not suffice. Simply adopting the design practices of foreign companies will doom U.S. industry to perpetual follower status. Educating new designers and performing research relevant to the needs of industry will require both the development of new faculty and intellectual and financial support from the companies at the forefront of engineering design practice. New research is needed to enable U.S. industry, when it is ready and able to accept new design methods and tools, to leap ahead of competitors.

2

Designing for Competitive Advantage

Engineering design, as discussed in Chapter 1, is the fundamental determinant of both the speed and cost with which new and improved products are brought to market and the quality and performance of those products. Design excellence is thus the primary means by which a firm can improve its profitability and competitiveness.

Yet few U.S. firms have adopted either contemporary design practices or product realization processes, and there seems to be inadequate understanding of how to go about improving current design practice. This chapter outlines the necessary steps to improving design practice and cites sources of information that should assist in this process.

Members of the committee visited several U.S. firms that use engineering design as a way to achieve competitive advantage. Information obtained from these visits, together with the collective experience of the committee, suggests that designing for competitive advantage requires much more than the adoption and use of new design practices. Firms that utilize design most effectively were found to:

- commit to continuous improvement;
- follow a product realization process tailored to their products;[15]
- use a set of design practices chosen to implement their PRP; and
- foster a supportive design environment.

CORPORATE COMMITMENT AND ACTION

Though many U.S. companies doubt their ability to win the competitive battles they are waging, a few have recognized the challenge that faced them and successfully fended off foreign assaults on their profitability and market share. What these companies have in common is recognition and acknowledgement of the potential or real threat to their market share and a shared corporate resolve to change the internal corporate culture in response to that challenge. Companies such as Xerox, Hewlett-Packard, and Ford, among others, have changed their internal cultures and reshaped the way they do business. From these and other companies, the committee learned that the first and most important step in introducing improved design practice is to generate corporate awareness of the leverage design can provide and the need for change to utilize that leverage. Change must begin with recognition of the importance and impact of design deficiencies and knowledge of possible routes to improvement. The committee's interviews and the collective experience of its members suggest that denial that a problem exists is the major obstacle to the introduction of new design processes and methods.

Denial is particularly prevalent in industries not yet besieged by significant foreign competitors. Until they have faced competitors that use superior engineering design practices, companies rarely recognize the advantages to be gained by improving their own design practices. Thus, many companies begin to improve their design practices only after they have lost significant market share to competitors that made such improvements years ago. Years of playing catch-up could be avoided, and competitive advantage gained, if enlightened management committed to continuous improvement under a PRP[16] in anticipation of rather than as a result of competition.

Businesses that have successfully incorporated state-of-the-art design practices have done so in an all-enterprise way. They have recognized engineering design as a vital part of their product delivery capability rather than as just another department in the company. This view ultimately required them to change many parts of the company beyond the design department; indeed, it usually spawned a totally new way of doing business.

Once a company recognizes the need to improve design, it must begin to identify solutions. Since deficiencies are rooted in organization, technique, and infrastructure, the main avenues of response are reorganization, adoption of formalized product realization processes, and involvement in research and education. In companies that successfully design for competitive advantage, the degree of external and internal change is often striking, reflecting a degree of self-examination rarely seen outside crisis situations. Successful programs of change typically feature strong top management leadership in setting corporate goals for improved design, development of metrics to measure progress toward these goals, creation of corporate centers of design excellence, extensive training programs for new hires and experienced engineers, and effective relationships with universities for research and technology transfer.

Knowledgeable observers point out that real change cannot be accomplished in a large organization without the impetus of a change agent, a group or department whose sole responsibility is to initiate change. Change agents are necessary because people whose main responsibilities lie elsewhere usually have neither the dedication nor the time to initiate significant change themselves. Xerox has assigned approximately 300 people (out of a corporate total of 113,000) to change-agent roles, Hewlett-Packard, about 1,000 (out of 89,000). Education and training programs, supported by senior corporate leadership and applied at the enterprise level, are effective and necessary supports for the change agents.

Support for change must include (1) programs to determine which practices worldwide would be most useful to the firm, (2) methods for securing support for the introduction of new practices, and (3) coordination of the change throughout the firm. Designers must be made a part of the change team, and the engineering design methods introduced must be explained as part of an evolving whole rather than as a series of unrelated fads. Unless engineers are educated in the value, goals, and necessity of a change plan, they will continue to use demonstrably inferior design practices. Because changing the product realization process affects the entire company, all employees, not just engineers, must be made part of the change process.

Though discussion to this point has targeted practice in large companies, much of the design and manufacturing in the United States is conducted in small and medium-sized companies (i.e., 500 or fewer employees) that often cannot afford extensive training programs or even separate design departments. Nevertheless, all of the principles stated here apply in and are crucial to the success and competitive position of smaller companies as well. Indeed, the integration and cross-communication implied in the product realization process may be more readily accomplished in smaller operations. Firms that cannot afford to conduct actions such as extensive training courses in-house can avail themselves of external courses and workshops. Large companies' training programs, for example, are often open to their suppliers.

THE PRODUCT REALIZATION PROCESS

Companies that design successfully have carefully crafted product realization processes that extend over all phases of product development from initial planning to customer follow-up. The PRP is their plan for continuous improvement. The decision to develop and operate under a PRP is a corporate one. Successful operation of a PRP requires extensive cooperation among a firm's marketing and sales, financial, design, and manufacturing organizations.

PRP's are not static, but evolve continuously. They change in response to feedback from production and incorporate new methods and tools. Design is an essential element of the PRP, and designers play a broad role in formulating and carrying out the steps of the PRP. The description that follows is an idealized composite of the various elements found in current processes, which vary from company to company.[17]

Definition of Customer Needs and Product Performance Requirements

A good product realization process begins with an exploration of business, marketing, and technical opportunities, followed by a firm definition of customer need and product performance requirements, including quality, reliability, durability, and other important factors such as aesthetics.[18] The new product's essential technologies are reviewed to ensure that inventions will not be required to produce it, and competitive products are analyzed to establish benchmarks for it.

Planning for Product Evolution

The technology review in the design phase indicates regions where technological advances or inventions can improve performance or reduce cost. In some industries, an entire range of products in the same line that require further invention, research, or development is mapped out, with planned evolution of features and capabilities, during this review. Core technologies for the future products are identified, and product performance specifications are defined with inputs from manufacturing, marketing, engineering, and finance.[19]

Planning for Design and Manufacturing

Cross-functional teams with representatives from marketing, design, manufacturing, finance, sales, and service are established. The design and evolution of manufacturing processes and production systems are projected. Necessary training programs are begun.

Product Design

The product is designed by the members of the cross-functional teams, including suppliers of purchased components, whose differing objectives are expected to balance one another.[20] The engineering effort aims at achieving a design that will exhibit little performance variation despite wide variation in the operating environment, product parameters, or even customer errors.[21,22] Simplification and standardization are applied to reduce the number and variety of parts and to make the product easily manufacturable. Conscious attention is paid to interfaces within the product and its manufacturing process and to the designer's planned evolution to the next model.

Manufacturing Process Design

The cross-functional teams establish requirements for product fabrication, assembly, and testing. They analyze tolerances, estimate costs, identify the best processing methods, plan assembly and test sequences, lay out the factory, and determine training requirements for factory personnel. All processes, manual and automated, are studied to determine whether they can consistently deliver products that meet specifications for quality, reliability, durability, and other attributes. Specifications are set for acquisition of in-process data needed to evaluate design and quality. Suppliers of manufacturing equipment are brought into the design process early to help define as accurately as possible the capabilities of any new machine or process that is to be used. Layout, production plan, and logistics for the factory and its suppliers are designed for minimum inventory and high flexibility.

Production

Statistical process control and in-process checks are used continuously. Inputs from these measures and observations from manufacturing personnel are continuously fed back to improve both the manufacturing and design processes and to aid in planning follow-on products.

Difficulties in the Design of Complex Products

In the foregoing idealized account of the product realization process, everyone cooperates, desired quality is achieved, and the product succeeds in the marketplace. In practice, the process is difficult and full of conflict and risk. Converting a concept into a complex, multitechnology product involves many steps of refinement. The design process requires a great deal of analysis, investigation of basic physical processes, experimental verification, complex tradeoffs between conflicting elements, and difficult decisions. For example, there may be insufficient space for a desired function unless costly development is undertaken, or space is taken from another function, affecting quality, fabrication yields, or ease of assembly. The original concept may not function as planned, and additional work may be required, affecting the schedule or requiring a change in specifications.[23] Satisfying the different and conflicting needs of function, manufacturing, use, and support requires a great deal of knowledge and skill.

IMPORTANT CONTEMPORARY DESIGN PRACTICES

If the product realization process is a firm's strategy for continuous improvement, design practices are its tactics. Most advanced engineering design practices are not particularly complex or difficult to understand and use. Indeed, many are becoming accessible in computer software packages, short courses, and books. Confusion exists because there are so many practices, with different, and sometimes overlapping, functions. Some (e.g., Taguchi methods) cover more than one practice. Because not all practices are applicable to or useful in the design of a given product, each company must carefully identify a set appropriate to its uses and incorporate them into its PRP. Companies must also establish means of assimilating new practices as they are developed. As mentioned elsewhere in this report, in order to leap ahead of competitors, companies must continually develop (or work with others who are developing) new practices to meet changing needs.

The following sections describe design practices under the headings of Traditional Practices, Modern Practices for Setting Strategy and Specifications, and Modern Practices for Executing Designs. The report does not attempt to discuss all important current practices, but rather to give the reader a flavor of the types of practices employed in the various phases of the PRP and to illustrate the great breadth of design activities.

Traditional Practices

The following traditional practices remain important and continue to evolve.

• *Searching and studying patents and the literature.* Patents and the literature, an extremely fruitful source of information generated by inventors, researchers, and other practitioners, can help designers avoid wasting time and money on approaches that won't work. Return per dollar of engineering effort invested is probably as great for patent and literature search and study as for any engineering activity. But because it is not recognized as a mainstream design activity and management fails to adequately motivate it, many designers shun this work. Consequently, the practice is underutilized in the United States. Efforts to review foreign literature are especially meager. In contrast, some Japanese firms assign engineers to this specific task; purchases of rights under U.S. patents are among Japanese firms' most effective investments.

• *Using standards of all types, as for components, procedures, computer-aided-engineering/design (CAE/D).* Use of standards can save design time, reduce uncertainty in performance, and improve product quality and reliability. It can also lead to economies of scale. Companies often define standard component lists and procedures with the goal of obtaining these advantages and then fail to enforce their use. New designers, failing to recognize the advantages of standards, tend to choose parts from their own knowledge or from the most familiar or convenient catalog. Unless a firm establishes standards and makes their importance known, any benefits that might result from their use will almost certainly be foregone.

• *Setting tolerances and the methods for checking them.* Greater understanding of physical factors that contribute to variations in controlled parameters and excellent metrology tools are powerful aids to designers in setting and checking tolerances. There is nevertheless a pressing need to better understand relationships between design tolerances and product quality and cost. Designers must have information and supporting tools to choose appropriate, cost-effective, and robust methods. Research topic A.4 in Chapter 4 describes research that will provide the requisite tools and information. A reference for an annotated, up-to-date set of references on tolerancing and metrology is included with the bibliography.

• *Prototyping.* Prototyping is an important tool for reducing time-to-market and providing models used to evaluate quality and producibility. In the past, prototyping proceeded through trial and error methods that were slow and cumbersome. At present, prototypes that are faithful representations of the final product are frequently required for use in experiments to optimize the product and work out assembly procedures. It is highly desirable to make these models with the same labor force and on the same line that the product will be produced on. However, this is not always possible, so better means of providing models are needed. Topic B.2 in the research agenda discusses the research necessary to create these models, and topic B.1 discusses research that will make it possible to expand the use of computer simulations, rather than actual physical models, as prototypes.

• *Analytical models: Both conventional analytical models and correlational models derived from design histories are powerful aids to engineering design and continue to evolve.* Correlational models, which relate design variables to performance measures using empirical data, are valuable tools in complex and incompletely modeled situations. The use of such models is illustrated in Figure 7.

• *Utilizing design reviews.* Although they are time-consuming and expensive and take reviewers away from their own projects, peer design reviews are immensely helpful in finding and avoiding faults and suggesting alternative approaches. For design reviews to be effective, management must motivate designers to participate and reward them for doing so.

Modern Practices for Setting Strategy and Specifications

New practices that have emerged to support the PRP are variously used to provide estimates of the cost and quality of new or redesigned products, in strategic evaluation of a firm's position relative to its competitors, in negotiations among the various contributors to a design, and even in negotiating with vendors and customers.

• *Product quality-cost models.* Models that give the designer the means of evaluating product quality and cost in the design phase are essential,

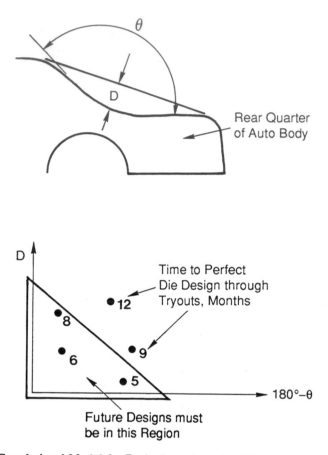

Figure 7: Correlational Model for Designing a Stamping Die

The rear quarter panel of a car is stamped from flat sheet metal. Designing the stamping die for the panel is very difficult because of the saddle-curve shape of the panel. The die must be redesigned and tried out several times, a process that takes many months and lengthens the entire car design process. To reduce die tryout time, one auto company identified two key design parameters and studied their influence by plotting tryout time for many previous designs. (Each dot on the plot represents a single design, the numbers the recorded tryout times.) The study showed that keeping the combination of these two parameters within specified limits would keep tryout time less than 8 months. The relationship is purely empirical. This guideline was given to the car body designers.

inasmuch as 70 percent or more of product cost is committed early in the design phase. New accounting methods, such as activity-based costing, provide accurate data on previous designs that can be used to generate quality-cost models, which are rapidly finding application in the design of both products and processes.[24] The research described under topic C.1 in Chapter 4 responds to this need.

• *Competitive benchmarking and quality function deployment.* The most successful firms benchmark continuously not only their own product performance and features, design tools and techniques, technology, production approach, and facilities, but also those of their most successful competitors. Reverse engineering is often a part of this activity. Quality function deployment (QFD)[25] is a process that seeks to ensure that products not only are technically correct and manufacturable, but also reflect customer needs. In QFD, an interfunctional team identifies product attributes consistent with customer needs and ranks them in an order determined by the customer. An appropriate weight is assigned to each attribute, and the attributes are converted into measurable parameters. The team then benchmarks these characteristics against the competition, chooses and incorporates in its own designs the best of what others have done, and develops only those features that provide competitive advantage. QFD is used by AT&T, Digital Equipment Corp., Ford, Hewlett Packard, IBM, and Xerox, among other companies.

• *Metrics for evaluating design practice.* Generating metrics to judge a design can produce useful feedback, both during a design and when reviewing earlier designs. Metrics are extremely difficult to craft, and the search for better ones, such as the number of engineering change orders or warranty costs, continues. Hewlett-Packard uses a metric based on "break-even time" (BET) to guide and evaluate product realization projects. The BET is defined as the time at which net operating profit (sales less cost of sales) equals total cost of design and development (TC). (See Figure 8.)

• *The "S" curve.* Almost all products follow an "S-shaped" life cycle curve. A product progresses from a stage in which its contribution is much greater than the cost of keeping it viable to a state in which an ever-increasing investment of engineering effort and capital are required to keep it in the market. It is important to know where each of a firm's products and each competing product are in their life cycles in order to gauge when to move to a new technology or approach with further growth potential. Some companies test the viability of their products by establishing teams that play the part of competitors with products on or approaching an "S" curve with a growth rate that surpasses that of the firm's own product.[26]

• *New management accounting systems.* Design's leverage derives from the fact that it determines product quality, cost, and time to market. For complex products especially, design is often a substantial fraction of total product cost. Because most companies cannot determine the contribution of

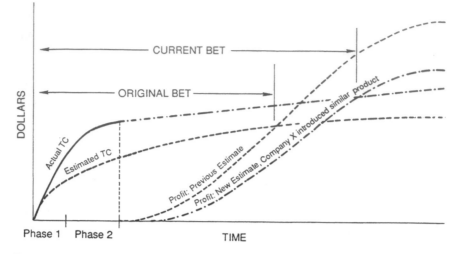

Figure 8: Break-even Time Metric

In this figure, the BET for a product is adjusted to account for increased design and development costs and the introduction of a similar product by a competitor. This illustrates the ability of the metric to estimate the effects of both internal and external events.

design to profitability, track design improvement, or effectively compare different product and process designs, the R&D budget for design and process development is usually determined by applying an R&D-to-sales ratio "about right for this industry" or "about equal to what we think our best competitor spends."

This situation results from the use of cost accounting systems, originally designed for other purposes, that provide only delayed and aggregated data perhaps based on labor or material costs,[27] and from the fact that R&D costs, being charged when incurred, are not associated with any product or process.

New methods that use detailed real-time information, obtained product by product, sometimes through computer-integrated manufacturing (CIM) systems, to provide required information at an affordable cost apply to design as well as to manufacturing. They can operate as overlays on existing processes and so need not supplant traditional cost accounting systems initially. It seems clear that these methods will eventually be widely employed to provide the data used to control production and track products, and also in design, to:

• determine the contribution of design to profitability,

- identify avenues to design improvement,
- establish product life cycle costs,
- provide accurate information for budgeting and planning new products, and
- document savings that result from reducing transactions.

These methods can also be used to generate, and associate with process design efforts, important nonfinancial measures such as quality, number of transactions, and manufacturing cycle efficiency.

These new methods are of two generic types: (1) operational control and performance measuring systems that use broad-based real-time data from production, and (2) activity-based costing methods that associate engineering and marketing costs, as well as labor, materials, overhead, energy, and machine and process time, with individual products.

In summary, new accounting methods make it possible to determine the contribution of product and process design to quality and profitability, to make intelligent allocations for R&D, and to determine explicitly the contribution of individual designers for purposes of recognition and compensation. Research aimed at creating and improving such methods is discussed in topic C.1 of the Research Agenda in Chapter 4.

- *The quality-loss function.* Taguchi defines quality in terms of quality loss: "Quality is measured by total loss to society due to fundamental variations and harmful side effects resulting from the manufacture and use of a product." Working from this definition, he introduces a quality-loss function[28] (qlf) to replace conventional go, no go specifications. Because it varies smoothly and continuously as a product parameter varies from specification, the qlf carries more information and hence is more useful than go, no-go specifications. By providing a common cost measure, it facilitates interactions between divisions in vertically integrated companies, between vendors and suppliers, and in resolving conflicts that arise from varying definitions of quality within marketing, manufacturing, and design.[29] In a typical application, the customer quantifies and supplies to the vendor the costs of departures from nominal specifications. The designer can then optimize these parameters and know what the customer is willing to pay for them. Both win. AT&T uses this concept to obtain agreement on transfer costs between divisions, and Texas Instruments' *cost-of-ownership* approach, used in working out integrated circuits supply contracts, is based on it.

Motorola's 6σ(six sigma) approach,[30] a derivative of the qlf and of Taguchi's robust design methodology, mandates designs that yield components that operate satisfactorily within $\pm6\sigma$ from the mean specified by the customer. This means that the product will exhibit only about 3.4 defects per million if the process mean shifts by 1.5σ in either direction. For example, 3.4 ppm defective means that the throughput from a process that uses 300 such parts and has 500 such assembly operations is about 99.73 percent (see Figure 5). Products produced thus are considerably more tolerant in the customer's application.

Modern Practices for Executing Designs

• *CAD and CAE.* Computer-aided design (CAD) and computer-aided engineering (CAE) have evolved over a period of 20 years into powerful tools that provide the ability to design mechanical, electronic, and architectural objects on a computer screen and transfer the design to manufacturing processes. In some cases, particularly for electronic objects, this transfer is seamless and entirely computerized. However, the use of CAD and CAE in U.S. industry, apart from electronics design, is surprisingly limited, and in even fewer cases is the output of the CAD system directly linked to computer-aided manufacturing systems or numerically controlled tools. A manager at one large automotive manufacturer estimated that only one-third of the company's designers used conventional CAD, and only a tiny fraction of those used three-dimensional solid modeling.

The capabilities of CAD and CAE systems do not meet the needs of many designers. Most often, the systems are used as little more than electric pencils that enable superior graphic presentation of designs. Only a few emerging systems permit any mathematical analysis to be performed on designed objects, other than in the well-established areas of finite element analysis and electronic circuit simulation. Methods are needed to link designs of interactive individual parts for purposes such as establishing tolerances or performing assembly analysis. Similarly, methods are needed to link product designs to other kinds of business data, such as inventory control, cost predictions, and factory modeling. Many of the topics discussed in the research agenda—for example, those involving design knowledge (A.2), computer representations of in-process designs (A.1), cost-quality models (C.1), tolerance synthesis (A.4), and design for X (B.3)—could greatly enhance this high-leverage area of engineering design.

• *DF(X).* These techniques, in which DF stands for Design For and X can stand for almost any operation (e.g., manufacture, assembly, test), are ubiquitous.[31] General Electric, for example, has an excellent program for design for the use of plastics that helps designers decide which type of plastic material to use in a given application. DF(X) techniques capture, in a standard procedure, all of the factors known to be important in a particular design activity. In the usual instances, costs are evaluated at each stage and at each interaction. These programs often provide examples and incorporate guidelines that help keep costs in the forefront, encourage the use of experience and standards, and prevent oversights. Though these programs are often specialized within a firm, progress is being made on generic methods of design for assembly and design for manufacture. DF(X) techniques are continuing to evolve, and new ones continually appear. New, improved techniques that can be expected from research are discussed in topic B.3 in Chapter 4.

• *Design rating systems.* Design rating systems such as those devised by Boothroyd and GE-Hitachi provide an impetus for design simplification and a method for tracking improvement.[32] These methods count parts of various types, promote the use of standard parts and the reuse of parts and subassemblies, and classify the motions required in assembly to provide estimates of quality and manufacturability.

A number of companies use design cost evaluation systems to compute the costs of capital expenses required by competing designs in order to obtain more realistic comparisons. Though neither perfect nor foolproof, such systems intelligently applied can reduce risks in cost, schedule, and design time. AT&T uses a computer-based system that evaluates designs transmitted electronically to a manufacturing facility by designers at 14 remote locations and flags designs that cannot be manufactured without manual intervention. Within one year of operation, more than 99 percent of the designs received by the system did not require manual intervention. The impetus for improvement is clear.

• *Concurrent design.* Concurrent design involves product designers, manufacturing engineers, and representatives of purchasing, marketing, and field service in the early stages of design in order to reduce cycle time and improve manufacturability.[33] This practice helps resolve what is sometimes called the designer's dilemma the fact that most of product cost, quality, and manufacturability are committed very early in design before more detailed information has been developed. Assembling a multidisciplinary design team permits pertinent knowledge to be brought to bear before individuals become wedded to their approach and much of the design cost has been invested. Differences are more easily reconciled early in design, and reductions in design cycle time that result from the use of this method invariably reduce total product cost. Though the use of concurrent design concepts has met with success, little is known about how to organize and manage concurrent processes and cross-functional teams effectively. Research that can enhance these methods is discussed in research topic C.2 in Chapter 4.

• *Simplification.* Simplifying a product by reducing the number and variety of parts and interfaces is often extraordinarily effective in reducing cost and improving quality and manufacturability. IBM's Proprinter development, General Electric's redesign of its electrical distribution and control product line for CIM production,[34] and Cincinnati Milacron's redesign of its plastic injection molding machines[35] are well-known examples of projects that applied simplification effectively. Reduction in the number of interfaces between parts and processes, a facet of simplification that is often overlooked, has proved to be particularly fruitful for AT&T. Simplification, though not difficult, is another nontraditional activity that must be made a specific design goal to be used to advantage.

• *Incremental improvement.* This technique builds on accumulated experience and developing technology to reduce product cost and improve quality. Warranty costs and experience from field returns are continuously monitored for opportunities for improvement. Technology is monitored to find particular parts or subassemblies that can be replaced with lower-cost, more reliable ones. Often simplification is applied. An incrementally improved product can usually be introduced to the market more quickly and with less risk than a new design. The successive stages of incremental improvement are readily discernible in the development of videocassette recorders, compact disc players, and cameras by Japanese firms.

• *Robust design.* Robust design is a systematic three-stage process, pioneered by Taguchi, to optimize a product or process. It calls for designers to examine all possible ways of eliminating quality loss in order to find the most economical one.[36] Following this protocol,[37] design commences with a systems design phase in which required features and function, including materials, parts, and tentative product parameter values, are selected. In the next phase, called parameter design, the designer systematically studies all parts to determine which do not significantly affect reliability or manufacturability. For these, the designer seeks low cost, commercial grade parts. For example, a punched part might be specified rather than a machined one or a ±20 percent resistor rather than one of higher precision. In the third phase, called tolerance design, the designer determines the tolerances required for the remaining parts to provide the broadest possible margins in manufacturing and operation. Because the number of parts is now fewer, more detailed analysis of the sensitivities of the design to parameter variation due to aging, environment, etc., can be performed. Often, cost-performance tradeoffs can be made specific.[38] A variety of tools can be used to facilitate this analysis. For electronic circuits, the group of programs generally referred to as SPICE[39] permits designers to optimize circuit operating margins given real or assumed statistical descriptions of component values and operating conditions. AT&T has equivalent mechanical and electronic design programs.

The SPICE program was developed in 1970 at the University of California at Berkeley by a team under Professor Donald Pedersen. It has been enormously successful, and many companies now offer customized versions of it. It is public domain software, and copies of the current version, SPICE 3D-2, are available from Professor Pedersen's group at the University of California at Berkeley for a nominal fee.

• *Use of designed experiments in the design of products and processes.* The application of appropriately designed experiments is useful for determining the relative importance of many different factors to reliability or process yield. Experiments can be constructed to use all of the experimental data in several ways, often reducing by orders of magnitude the amount of experimental data that must be collected compared to the traditional approach of varying one parameter at a time while holding all others fixed. The experimental approach is most useful when the number of variables is large, the effects are relatively substantial, and interactions among the various parameters are unknown. Use of this approach is rapidly being made easier by the availability of good personal computer-based software tools.[40]

The methods used in this approach were developed by R.A. Fisher, who applied them in agricultural experiments in England during the 1920s.[41] Professor G. E. P. Box and others extended Fisher's methods and applied them in many industrial applications. Professor Box's approach is straightforward and satisfactory for most problems.[42] Dr. G. Taguchi has promoted and applied these methods in design and troubleshooting.[43] Japanese and U.S. automobile industries use these techniques extensively, sometimes performing tens or even hundreds of experiments during various stages, particularly early stages, of product design.

UNDERSTANDING, MOTIVATING, AND SUPPORTING THE DESIGNER

The design of products and processes is a creative activity that depends on human capabilities not easily measured or predicted. The most effective designs are acts of creativity that rank with those in the fine arts. We are not within sight of the time when machines can perform the design function, though tools can certainly aid the designer. Dependency on designers makes it vitally important that companies understand the nature of the design task and the nature, characteristics, and needs of people who design effectively in order to be able to create an environment that stimulates and nurtures them.

The design environment is set in large part by the organization of and strategy for design. A formal, well-supported product realization process can make an important contribution to a productive design environment. In the following sections, we discuss briefly the nature of the design job and the designer, and some steps that can be taken to provide a supportive design environment.

The Design Task

The design task, which once could have been adequately defined as achieving a function at a specified cost, has broadened under competitive pressures to include at least three broad areas of endeavor:

- designing products and processes to meet many constraints;
- developing and improving design tools and processes, including the PRP; and
- standardizing parts and generating specifications.

The various practices described previously represent only a part of the designer's task in designing products and processes. Table 1 lists some of the factors besides quality, cost, and time to market that can make or break a design.

In addition to designing products and processes to meet many constraints, designers often have the task of integrating numerous separate procedures into complete processes. This function, which may include some tool development, controls the flexibility of the resulting process and the time required to execute a design. For example, a designer may develop a computer program to link the output of a CAD system to an automatic parts insertion machine, eliminating manual data transfer, and thus saving time and reducing errors.

Designers also work with other parts of the firm, with customers, and with vendors as they use many of the tools and techniques described earlier to obtain the information needed to set product specifications. As the principal agents for the PRP, they must have strong interpersonal skills as well as sound technical skills and creative ability.

The breadth of knowledge required by the practicing engineer today is enormous, encompassing many topics not emphasized or included in standard engineering curricula. Dr. Joel Spira has developed a detailed outline for a course that addresses issues that the practicing engineer will undoubtedly encounter in today's environment. A course based on this outline will be given at Cornell University, and similar courses are being considered at other universities. Dr. Spira's outline[44] is included as Appendix B.

The Designer

Who designs? In most firms today, design is not limited to those who are educated as designers or who spend most of their time designing products or processes. Many more engineers and scientists participate in design than those whose job assignments are design. An increasing number of people are involved in activities, such as competitive benchmarking and reverse engineering, that are more analytic than synthetic in nature. To derive information useful to the designer, these people must understand design. Those who do process design and systems integration must also have knowledge of the design process, as must the many engineers and scientists who work on CIM or the PRP. Manufacturing engineers who work on teams with designers and marketing people must understand design as thoroughly as manufacturing in order to arrive at manufacturable products.

TABLE 1 Touchstones for Design

In addition to quality, cost, and time-to-market, all of the following considerations are important in design:

Customer—Who is the customer? What does he or she really need?

Stakeholders—Understand the positions of those who have stakes in the product's success or the status quo.

Ease-of-use—Human factors design needs to be addressed early in the process.

Documentation—Essential; match to user's needs; start early.

Cultural change—If development or production of this product or process requires cultural change, its introduction will not be easy or swift.

Patent/Copyright—Plan for this early to avoid pitfalls and to get high quality coverage.

Legal/Regulatory—Consider early. Such obstacles have delayed or damaged many projects.

Environmental Impact—Determine if the manufacture or use of any product may adversely affect the environment.

Manufacturability—Has the manufacturing engineer been on the team?

Aesthetics—These hard-to-define characteristics are also critical.

Dynamics—How does the product or process behave in non-steady state conditions?

Testability—How will the product be tested? Where, by whom, at what cost?

Prototypes—Consider how the final product may differ from the prototype if prototype and production processes are not identical.

Universality—Universal solutions almost never work.

Simplicity—Strive for beautiful, simple designs. They often work well.

Appearance—If the design doesn't look right, watch out!

Interfaces—Many otherwise sound designs fail because of unanticipated problems at interfaces.

Maturity—Where is the product on its "S-curve"? Is it time to jump to a new approach?

Partitioning—Consider partitioning to provide additional degrees of freedom.

Models—Do the mathematical models used in design apply over the anticipated range of use?

Scale-up—Do not undertake this lightly. Proceed by small increments.

Transportation—What happens to the product in transportation?

It was often noted to the committee that individuals who enjoy design and excel at it take a fundamentally different approach to their work than engineers and scientists whose forte is analysis. In summarizing observations about the attributes of excellent designers, the committee recognizes that generalizations about human capabilities are subjective and may have important exceptions. Nevertheless, this summary may be useful in helping engineering and human resource managers recognize, support, and reward people with strong design abilities.

First, effective designers seem to have great associative power that lets them recognize and draw upon parallels in other fields for ideas. Consistent with this is the observation that such individuals usually have eclectic interests and often roam far afield in science and engineering. Many have a strong interest in puzzles and games that involve numerous permutations. Above all, such people were said to be "interested in everything."[45]

Second, good designers presented with a problem always seem to respond with a flood of ideas rather than a single solution. When asked, they often reply that this is part of the way they think and that they are often not very good at sorting out alternatives. Instead, they look to interactions with associates to sort out the good from the bad and, in most cases, to complete the formulation of the good solutions.

Third, good designers often have strong inner-directed personalities. Being sure of their own worth and contribution, they are able to accept with equanimity the guffaws at the poor solutions they propose along with the kudos for the good ones.

Fourth, the output of designers spans a very broad range. A number of people expressed the belief that a large fraction of the design in their firms is done by a small fraction of the most effective designers. They suggested that the range of output of designers is so great that it, like many other human attributes, is best expressed on a logarithmic scale.

Finding, Supporting, and Rewarding Effective Designers

Some committee members and some people in the firms visited believe that the most effective designers have, in addition to analytical ability, the same sort of strong "right brain" skills as artists and poets—that it is this sort of associative skill that lets them come up with ingenious solutions. If this is the case, industry can certainly do a better job of finding and placing people with these abilities. Current employment screens that rely almost entirely on measurements of analytical and logical skills probably misdirect many people who have strong associative abilities and lesser, though adequate, analytical skills. If it is accepted that the most productive designers come from the pool of people with this associative talent, it follows that efforts to identify and select people with the needed innate talents should pay off.

It is also reasonable to expect that design skills, like other human skills, can be stimulated and honed through study and practice. Yet, in contrast to the large body of literature on scientific method, there is relatively little material on what we call by analogy "design method," or the "mental discipline of design"—that is, how one goes about finding solutions. A couple of notable exceptions are the work of E. Bright Wilson,[46] whose guidance to scientists in designing apparatus and experiments is as useful to engineers as to those in the physical sciences, and Polya's books,[47] which are almost unique in providing guidelines about the thought processes, as opposed to the technical details, of design. A book by Henry Petroski,[48] written not just for technical audiences, addresses the sharpening of design skills through a study of what is to be learned from design failures.

Though engineers who design have long understood the great leverage their work exerts on product quality, cost, and time to market, they believe that this leverage is not widely appreciated and that they are bound up in a maze of constraints that have little to do with product realization. Many are concerned that management fails to distinguish their high-level work, which requires innovation and the use of the most sophisticated analytical tools, from that of the draftsman, who is often dubbed a designer in an attempt to enhance the image of drafting. One of the most effective actions management can take is to understand and to acknowledge the designer's role in the business. Most companies that use design effectively include specific recognition of design excellence as a part of their product realization processes. Some firms now have regularly scheduled programs at which the engineering designers describe their products and processes and are recognized for their contributions.

To help design engineers maintain their proficiency, alertness, and knowledge, firms must emphasize continuing education through high-level commitment and operational priority at all levels. Continuing education takes many forms, both within and outside the firm, and participation should be regarded as a vital, continuing part of the design task. This behavior is characteristic of "best-practice" firms. A full treatment of this subject is given in a 1988 NAE report.[49]

Finally, adoption of new management accounting methods, such as activity-based costing, can help designers improve designs of products and processes, and help to insure that designers' contributions are adequately recognized.

Elements of a Supportive Design Environment

The following elements, found in companies that utilize design effectively, contribute to designer efficiency and productivity.

- A coordinated companywide approach to product realization that explicitly recognizes the designer's role
- The availability of continuing education (absolutely essential both for learning to use new tools and techniques and as a source of stimulation)
- Easy access to the literature and strong encouragement to use it
- Involvement with research in the interest of making it relevant to industry practice and to speed transfer of research results into practice
- Recognition for achievement in design (for many, more important than salary)

- Rewards for all tasks performed, including nontraditional ones for designers, such as working with vendors and working with customers' engineers as a part of the sales effort
- An active role for designers in choosing the tools and methods a company will use and in formulating the company's product realization process.

Summary

Clearly, the first step to improving the practice of engineering design in the United States is for industry to acknowledge the need for improvement. This done, industry must take the initiative in learning about and adopting appropriate best engineering practices, undertaking and collaborating with universities on relevant design-related research, and encouraging the academic sector to rethink undergraduate and graduate design education.

Companies must review contemporary design practices in light of their product realization processes (or lack thereof) and must attend to the nature of the design task and the designer and build a supportive design environment. The resources to do these things are largely within the firm and largely organizational.

3
Improving Engineering
Design Education

Engineering education in the United States has undergone many important changes since World War II, leading to impressive improvements in the engineering graduate's knowledge of the engineering sciences, mathematics, and analytical techniques. These changes include restructuring to emphasize the engineering sciences as a coherent body of knowledge, the introduction of new disciplines, the creation of an extensive system of research and graduate programs, and the partial integration of computers into curricula.

While these improvements were taking place, the state of engineering design education was steadily deteriorating with the result that today's engineering graduates are poorly equipped to utilize their scientific, mathematical, and analytical knowledge in the design of components, processes, and systems. Strengthening engineering design education is critical to the long-term development of engineers who are equipped to become good designers and leaders and who will provide a lasting foundation for U.S. industry's international competitiveness.

Design is the characteristic activity of engineers, although many engineers are not involved directly in performing design functions themselves. One analysis of activity of engineers, shown in Table 2, shows 28 percent involved directly in development, including design; however, an understanding of design is required to work effectively in engineering management, production, technical sales, and other functions. The fundamentals and nature of design are not taught in courses devoted to engineering sciences, yet well-prepared graduates need such knowledge as they start their engineering careers. Consequently, design must be a significant component of undergraduate engineering education.

TABLE 2 Primary Activities of Employed Engineers

(Percent of Engineers)	
Research	5%
Development, including design	28
R&D management	9
Other management	19
Teaching	2
Production/inspection	17
Other and unreported	21

Source: National Research Council, Engineering Education and Practice in the United States: Foundations of Our Techno-Economic Future, 1983, p. 91.

THE GOALS OF ENGINEERING DESIGN EDUCATION

Undergraduate and graduate engineering education establish the foundation for successful design practice, design research, the teaching of engineering design, and careerlong learning. Undergraduate engineering programs seek to impart knowledge in basic sciences and mathematics, as well as fundamental knowledge and capabilities in engineering analysis and design. Graduate programs seek to build upon the undergraduate foundation and reinforce specialized knowledge and capabilities in engineering science, design practice, and research in engineering sciences and design.

Undergraduate Engineering Design Education

Undergraduate engineering design education must:

• show how the fundamental engineering science background is relevant to effective design;
• teach students what the design process entails and familiarize them with the basic tools of the process;
• demonstrate that design involves not just function but also producibility, cost, customer preference, and a variety of life cycle issues; and
• convey the importance of other subjects such as mathematics, economics, and manufacturing.

To achieve these goals, design must be distributed throughout the engineering curriculum, beginning with introductory design courses, which serve the dual purpose of introducing the design process and demonstrating the relevance of the engineering courses to design, and continuing as a part of the more advanced engineering courses. Additional material, such as probability and statistics, economic analysis, optimization methods, and manufacturing principles, is needed to understand modern engineering design and should be included in the engineering curriculum (not necessarily as discrete courses). The combination of engineering fundamentals and introductory design should culminate in senior design projects that apply the concepts learned to significant broad design problems. Recognizing that learning comes not from doing alone, but also from prompt evaluation and criticism, an essential ingredient of the senior project should be informed evaluation of and feedback on student work. Metrics for evaluating student design projects need to be developed to provide this feedback.

Although it is imperative that students spend some time in real industrial design settings, it is equally critical that on-campus laboratory facilities be provided to expose them not only to the functional aspects of design but also to production, quality control, testing, and so forth. Although students are not expected to become experts in these areas while in school, they need to develop a genuine awareness of their role and importance. Computational tools and specialized software are essential. Limited time during a one- or even two-term course necessitates a supportive environment to develop and validate ideas from concept to completion. This requires sufficient space and communication capability for project teams to work together effectively and sometimes to carry paper projects to physical realization.

Graduate Design Education

Graduate design education should be directed toward:

- developing competence in advanced design theory and methodology;
- familiarizing graduate students with state-of-the-art ideas in design, both from academic research and from worldwide industrial experience and research;
- providing students with working experience in design;
- immersing students in the entire spectrum of design considerations, preferably during industrial internships; and
- having students perform research in engineering design.

A continual stream of design-oriented doctoral graduates with new design knowledge is needed to supply faculty who can teach undergraduate engineering design. Other measures, such as faculty-industry exchanges and faculty retraining, though important, especially in the near term, cannot produce the permanent infrastructure change that is sorely needed. New doctoral graduates strong in design who will succeed in faculty careers after gaining industrial experience are required. Even graduates who do not intend to specialize in design need to understand design better in order to relate engineering science courses to practice.

Design-oriented graduate students who will later become faculty members will help to develop a larger and stronger constituency for design in academe. They will pursue research, thus generating more graduates, and upgrade the design portions of graduate engineering programs.[50]

THE STATUS OF ENGINEERING DESIGN EDUCATION

A 1985 report on engineering education[51] points out that most educational institutions that offer engineering programs have become one of two types since 1950, (1) research universities or institutions whose graduate and research programs are heavily dependent on contract research, and (2) colleges that have as their primary focus undergraduate education in engineering. Each type of institution grants approximately half of the baccalaureate degrees in engineering. The motivations and problems of these different types of institutions must be kept in mind when making generalizations about engineering education. Variations in emphasis and quality of educational programs, from program to program in a given institution as well as from one institution to another, must also be recognized. In light of this diversity, the following observations are presented.

Undergraduate Programs

Several recent reports and papers have pointed out deficiencies in design education and called for its strengthening.[52]

Employers of recent engineering graduates frequently commend many aspects of the graduates' performance, particularly their facility with analytical calculations and computers.[53] With the possible exception of writing and speaking, these employers find design to be the engineering graduates' most prominent weakness. Sometimes these complaints are voiced in terms of recent graduates "not understanding that costs are important" or "not realizing that someone has to make what they come up with and someone has to sell them" or "not realizing that this is a complex organization." The complaints may not use the term design, but they relate to knowledge that should be woven into the design parts of curricula. Complaints from industrial employers would be more strident if their expectations had not been lowered by years of neglect of this area by schools.

To learn engineering design takes longer than any university education can last. University teachers are painfully aware that they do not, and never will, have the time, knowledge, or facilities to teach engineering students everything they need to know about design before they begin their professional careers. Industrial companies are also generally aware of this, and those with the requisite resources are usually willing to take on part of the educational load, though many find that they must too often focus on remedial activities rather than on topics particular to their business.

Undergraduate engineering curricula are required by the Accreditation Board for Engineering and Technology (ABET) to meet these minimum quantitative criteria:

(25%)	Mathematics and Basic Sciences:	32 semester hours
(25%)	Engineering Sciences:	32 semester hours
(12.5%)	**Engineering Design:**	**16 semester hours**
(12.5%)	Humanities and Social Sciences:	16 semester hours

(These criteria total only 75 percent to allow for flexibility, and the broad definitions allow much latitude in emphasis and approach even within the curricular components listed.)

Even this minimal level of design emphasis is often not met by undergraduate curricula.[54] ABET annual reports show that deficiencies in engineering design are one of the leading causes of less-than-most-favorable accreditation actions. Each year 60 percent or more of the engineering programs evaluated receive less-than-most-favorable accreditation actions. Table 3 shows the major deficiencies found in such programs during three recent years. Among all the engineering programs evaluated by ABET in 1989, 33 percent were cited for deficiencies in engineering design.

It must be emphasized that these are deficiencies relative to the current ABET criteria that include no mention of concurrent engineering, total product life cycle, and experience working in a team. Current criteria also do not

TABLE 3 Deficiencies of Engineering Programs Receiving Less-Than-Most-Favorable Accreditation Action, 1987-89 (The percentages total more than 100 because programs were in most cases cited for more than one deficiency.)

Specific Deficiency	Percentage of Programs Cited		
	1987	1988	1989
Engineering Design	44	44	49
Laboratory Plan	50	33	34
Laboratory Equipment	34	30	30
Resource Allocation	36	34	27

Source: ABET Annual Reports, 1987, 1988, 1989.

require the inclusion of economic evaluations and consideration of alternative solutions.

Often institutions claim design content in engineering science courses. When those courses do involve design, the effect is productive, but examination often reveals that the claimed design content is not there. Courses devoted to design are often poor and reflect an antiquated view of the field. Few are multidisciplinary or present modern design methods, and even the customary senior design courses seldom treat the processes involved in sound contemporary design practice. Often, too much is expected of these senior design courses when prior courses have failed to provide sound preparation for them. When, for example, a senior design course is a student's only exposure to integrated design activities such as concurrent design, detailed consideration of alternatives and constraints, significant economic analyses, and working as part of a team, the experience is likely to be shallow.

To resolve current curriculum deficiencies, universities must comply more fully with the intent of ABET criteria and ABET must strengthen the design emphasis in its criteria.

Employers report that many recent engineering graduates have only a weak grasp of some of the curricular material they have studied. One reason may be that many students who study design tools such as engineering economy, statistics, probability, and various mathematical, numerical, and computer methods do not get an opportunity to use them in subsequent courses. The integration of course material needed to alleviate this problem requires time-consuming faculty cooperation and teamwork, but faculty incentives and rewards are based chiefly on other activities.

Although the advantages of interdisciplinary design teams are recognized in both industry and education, truly interdisciplinary teams are rare in design courses. Instead, in most engineering colleges design is fragmented, isolated by discipline, and uncoordinated. Organizational problems are one cause. Another is that interdisciplinary teams require greater teaching effort, which is usually not recognized in evaluating teaching loads. Yet another cause is the reluctance of faculty members to become involved in interdisciplinary activities. This is the same reluctance that gives rise to narrow courses such as "heat transfer design" and "control system design," which inspection usually shows to involve functional design almost exclusively, not even approaching the breadth of modern engineering design.

Although many excellent design courses are taught, adequate design education requires a coordinated approach among several courses in the curriculum. The following features characterize the few curricula that include strong comprehensive design programs.

• Design is taught in several courses throughout the curriculum, not just in "capstone" design courses in the final year.

- The final-year courses, at least, use broad, new, open-ended design problems.
- The design program as a whole covers many of the characteristics of design and a variety of design experiences. Ample attention is devoted to the generation and evaluation of alternative designs.
- The design courses are taught by several faculty members, many of whom have full-time industrial experience.
- There is close cooperation among faculty in integrating the courses.
- Students are closely guided in early stages of their design experience, and their work is carefully evaluated.
- Students gain some experience working in groups.

One characteristic not yet observed widely, even in otherwise strong programs, is consideration of formalized modern design methodologies in the required design courses. Although some faculties are beginning to incorporate this focus (and its growing importance suggests that it should be incorporated), it is not mentioned in current ABET criteria.

Engineering education does not now adequately prepare graduates to keep current with engineering advances throughout their professional practice. Many baccalaureate engineers have never read a current engineering paper or made an in-depth library search. Engineering managers report that most engineers in industry do not follow the refereed engineering journals. These observations clearly reveal undergraduate engineering program shortcomings that relate to design ability.

Few engineering graduates have been taught to expect continued learning to be part of their careers. In job interviews, they seldom ask prospective employers about formal continuing education opportunities, though this should be a primary factor in evaluating an employer. Although educators sometimes respond that employers should take the initiative in solving these problems, instilling in students a motivation for continuing their learning is clearly an educator's job.

Two recent reports have described the status of continuing education in engineering.[55] One of these, *Focus on the Future*, presents a plan of action that should be considered by every engineer and employer of engineers. Familiarity with these reports would probably help engineering faculty members in stimulating student expectations of careerlong continuing education.

Graduate Programs

Because one of the most valuable features of graduate study is the flexibility to offer programs, or "plans of study," tailored to each student, graduate programs cannot be structured to the extent that undergraduate ones are. The plans of study followed by most engineering graduate students have no design content whatsoever. When design is included, the quality of courses available and of design projects and design-related research varies widely. Finally, few graduate programs require students focusing on design to work in industry, though such experience is critical to learning about the realities of engineering design practice.

The inadequacy of the design component of undergraduate education ill prepares students for graduate design courses. This contrasts with the case for students in the engineering sciences, who have often taken undergraduate electives beyond the required courses. In addition, engineering graduate programs that admit students with nonengineering degrees, who are likely to have no prior training in design, force graduate courses into a remedial mode.

There are simply too few strong graduate programs focusing on modern design methodologies and research to produce the qualified graduates needed by both industry and academe. Limited funding for design research impairs the quality of graduate programs in design and reduces the number of graduate students for whom work in the field can be supported. Even the stronger programs rarely involve industry experience that would elucidate the realities of engineering design practice. Engineering design education cannot mature until strong graduate programs that focus on modern methodologies and the research needed to advance them begin to produce qualified graduates who are committed to design as a career.

Faculty

Most faculty members are neither trained to teach design nor cognizant of its importance. Significant improvements in engineering design education are highly unlikely without strong, knowledgeable, enthusiastic faculty members who interact with colleagues in their own departments, in other departments, at other institutions, and in industry. At present, the number of faculty who consider design part of their mission and responsibility is quite small.

The shortage of faculty to teach design is much more severe at the graduate than at the undergraduate level. One significant characteristic of this obstacle is the amount of time that will be required to overcome it even after actions are initiated. Inasmuch as recent developments in engineering design have rendered obsolete much of past practice in both industry and education, faculty not current in design will require a significant amount of study to become current.

Few faculty have significant industrial design experience or possess an understanding of manufacturing, and their contacts with industry (if any) are usually limited to consulting on nondesign issues rather than involving real-time design and manufacturing activities.[56] Industry does little either to support or guide university research or education in engineering design, and only rarely do designers from industry join university faculties. In addition, university and industry design researchers are often ignorant of each other's work.

In view of the rapid evolution of design, continuing education is as critical for faculty members as it is for engineers in industry. Although they may read current literature, such people seldom enter continuing education programs. Their professional skills could be enhanced by continuing education course material, as well as by increased interaction with engineers in industry. Currently, industry conducts much of the continuing education in design.

Although experienced teachers of modern design courses find the activity stimulating and rewarding, most engineering faculty members, being unfamiliar with design teaching, consider it difficult and do their best to avoid it. Others, familiar with the old, generally pedestrian, "cookbook" style of engineering design education and practice, perceive design to be an inferior enterprise. Emerging modern design methods and scientific foundations for design are slowly changing this view.

In contrast to the engineering sciences, design knowledge is diffused and poorly organized. Some of the most valuable knowledge is anecdotal and so diverse that a taxonomy of design domains is difficult to construct. Generally applicable design principles are only now beginning to appear. Design research results should add rigor and formalism, as usually seen in basic science and engineering science courses, to design courses. Case studies should offer a useful technique for teaching design, but, although instructive cases abound, most are not well enough documented to be readily useful to teachers. At best, preparation, including establishing the requisite industrial contacts, requires an extensive time commitment on the part of faculty.

Similarly, textbooks that provide a comprehensive insight into the field of engineering design are rare.[57] Other teaching materials, such as videotaped lectures, case studies, and software, are virtually nonexistent and need to be prepared by teachers. This problem is complicated by the domain-specificity of much of the material.

Most engineering faculty are unfamiliar with the many-faceted instructional techniques required for design education. Techniques for teaching analytical courses in a specialty are familiar and more narrowly defined; they involve classroom lectures and discussion of well-defined material that can be presented largely in isolation from other material. Design, in contrast, is multidimensional, and the hands-on conduct of the design process is not readily separated from other material. Furthermore, because of the multiplicity of design criteria, of which function is only one, and sometimes not the primary factor, design problems seldom have unique solutions. Considerable effort is required to guide students and evaluate their work when they or teams of them are following diverse paths, including some the instructor had not anticipated. It is not surprising that few faculty have proved to be effective teachers of courses that involve open-ended team projects in design.

Effective design teaching also faces institutional obstacles. The need for interdisciplinary design teams is recognized, but establishing design courses to include them calls for cooperation not just among faculty members but also among departments, which is even more difficult to achieve.

In a typical faculty reward system, especially in research-oriented institutions, tenure, promotion, and salary are based largely on publication of refereed scholarly journal articles and on grant and contract funding. Despite their obvious value, there is little reward for professional activities, work in industry, or even teaching and other efforts to improve the education of students. Teaching design is particularly time consuming and held in low regard by the academic community, particularly outside the engineering school. In this environment, in which research papers in academic journals are the principal measure of faculty achievement and capability, there are few channels for publishing engineering design achievements. Consequently, young faculty members logically and rationally conclude that design is a dangerous and unrewarding career focus. Experienced faculty members who have built reputations in the engineering sciences often want to work in those areas where they have more confidence in their ability to develop further their faculty credentials. The general success of engineering faculty members in teaching engineering sciences may even motivate academics to further emphasize engineering sciences in curricula at the expense of engineering design. Beyond encouraging faculty members to avoid design teaching themselves, the reward system may further influence them to reduce curricular emphasis on design in order to reduce the total amount of design teaching required of their departments.

IMPROVING DESIGN EDUCATION

The problems discussed thus far center on the failings of the curricula and on barriers to enthusiastic and effective faculty participation. The initiative for immediate improvement of design education and for laying the groundwork for its longer-term sustained improvement lies clearly with educational institutions. Even without additional resources or restructuring, significant improvement is possible simply by assuring that each engineering curriculum fully meets the letter and spirit of the current ABET criteria for undergraduate programs. Major improvement depends on major revisions in the goals and practices of educational institutions. Engineering design education is seriously deficient, and strong steps are needed to revitalize it.

Because universities are neither penalized if they fail to nor rewarded if they do support engineering design education and research, academic administrations and faculty feel no pressure to change. They often disclaim responsibility for the problem, blaming it on "the system." Inasmuch as these individuals accept the "system," it is they who must take the lead in changing it. Changing systems that do not work well is an important function of leaders.

. Professional engineering societies, largely through ABET, have often provided leadership in improving engineering education. (In the past, examples have been seen in connection with engineering sciences, humanities and social sciences, communication skills, laboratory facilities, and computer use.) The need is urgent for them to lead in improving the design part of engineering education. One of the recommendations of this report is for modernization of the ABET accreditation criteria, and professional societies that are ABET participating bodies must take the initiative in revising them. Also, it must be remembered that the selection and training of ABET program evaluators is chiefly the responsibility of the societies that are ABET participating bodies.

Professional engineering societies, through their education arms, should encourage the further education of design teachers and increase the awareness of all faculty members of the importance of engineering design. The guidance of practicing engineers is essential.

Institutional Initiatives for Reform

Industrial firms and educational institutions are so different in purpose, organization, and other respects that applying the experience of one to the problems of the other is seldom likely to succeed. However, it appears that improving the teaching of engineering design in universities may need to follow the same steps that successful programs of design improvement in industry have followed, namely:

- recognize deficiencies in design quality;
- exert strong, high-level leadership in establishing goals for improved design;
- develop metrics to measure progress toward these goals;
- create change agents to plan and implement improvements;
- establish extensive training programs for both new and experienced teachers.

Universities are frequently sensitive to, but unmoved by, criticism from outside. Their response to adverse criticism is usually to deny the problem and then to reiterate the arguments that support existing practices. If their primary metrics are numbers of faculty publications, dollar volume of research expenditures, and numbers of awards received by faculty members, they will expound on the relationship of these to educational objectives (and in publicly supported universities, to public service objectives). These relationships cannot be denied, but they do not justify the failure to address other metrics more directly indicative of the institution's success in meeting its educational goals.

Recognition within academia of deficiencies in design teaching will not be driven by loss of market share or financial results. It must begin with perceptive faculty members and administrators (or occasionally with outsiders) who can visualize or have observed strong design programs and recognize their importance. They must make clear to all involved that (1) the fundamentals of design are an essential part of engineering curricula, (2) additional engineering science courses cannot make up for a deficiency in design teaching, and (3) simply providing students with design "experience" is inadequate, because design fundamentals must be taught.

These initiators must overcome the conventional objections to curricular change. A familiar objection to adding material is that curricula are already overcrowded, although it is frequently pointed out in the engineering education literature that more effective integration among courses, greater use of new teaching technologies, and closer examination of material to identify requirements which can be eliminated are underutilized for updating engineering curricula. However, improving design education involves a willingness to try new approaches, increased faculty teamwork, and supplanting outdated approaches rather than adding new material. Curricula now strong in design have not compromised other components to provide this strength.

Only if strong leadership stresses the importance of design will faculty and administrators establish goals for improved design teaching. Input from industrial firms that are using modern product realization processes or concurrent engineering is essential in establishing goals. Faculty must be shown that contributions to setting and achieving these goals will be rewarded.

Metrics must be established to measure progress toward the goals. Operational metrics for design education are harder to establish than metrics such as enrollment, degree production, research funding, and number of faculty publications, but each institution must devise a suitable set and modify it on the basis of experience. The numerous instances of industrial firms that persisted in using the wrong metrics should be kept in mind. Existing metrics in universities are often in conflict with the metrics that indicate success in educating students. For example, hiring capable, vigorous senior engineers from industry has been suggested as a step toward solving the design faculty shortage. Such people are available, but they are unlikely to be hired by institutions whose primary metrics are research publications and securing of research funding, because young, inexperienced faculty members fresh from doctoral programs are more likely to contribute to these metrics, especially in view of the pressure on them to do so in order to secure tenure.

Academic institutional structures militate against reform, and although some problems could be solved with currently available resources, the lack of incentives for universities to change is an obstacle in itself. Many observations confirm that the designation of an agent to plan and implement change has been a key to improving design in industrial firms, which are also resistant to change. Because universities are perhaps even more reluctant to change than large industrial firms, the need for designated change agents may be even greater in academe. To ensure its effectiveness, the form of the change agent must be determined by each institution under alert, high-level leadership. An important function of this agent should be to promote interdepartmental activities and relations with industry. It can also provide an effective conduit to the national clearinghouse described below and a needed focus for design teachers who are often scattered across several departments. Though faculty may be reluctant to form another committee, best-practice companies have found dedicated functional change agents to be essential to implementing changes in established infrastructures.

External advisory boards, carefully appointed exclusively for the evaluation and improvement of design education, can provide effective guidance for on-campus change agents.

Aiding Teachers of Design

Even if universities were to change goals and institute rewards to make design teaching more attractive, there would still be a lack of adequate classroom and laboratory support materials, and most faculty would still require help to teach design effectively. Currently, no good source of information on design theory, methodology, and available tools is easily accessible to all teachers of engineering design. A national clearinghouse for design instructional materials could make the task of teaching design easier for many faculty. Such an organization would collect, compile, and disseminate information on design theory and successful industrial practice worldwide. Design research, teaching methods, and design software tools need to be reviewed and the results published. Though this information can be valuable for industrial practice and research, it is critical that it be cast in a form appropriate for faculty teaching use. Access could be provided through periodicals, on-line data bases, seminars, workshops, and trade shows, as well as through texts and problem manuals, whose commercial publication needs encouragement. In addition to accelerating the rate of information dissemination to schools and industry, the clearinghouse could also facilitate the introduction of standards and common representations (e.g., IGES, PDES).[58] The selection of design tools could be supported by publishing benchmarks and industrial experiences with such tools.

Participation in design education training and workshops, as well as opportunities for faculty to visit, observe, and participate in outstanding design courses, both in industry and academe, will further aid faculty in teaching design. Training programs for faculty are uncommon but are especially needed in design because the need for more faculty to contribute to the design component of engineering curricula is currently so great. If faculty members, inexperienced in design and design teaching as many are, are to be induced to engage seriously in design education, the task must be made easier for them. Training programs are one avenue. On-campus training programs can also assist by providing both teaching materials and convenient continuing access to new information.

Improving University-Industry Interaction in Design Education

Industrial firms employ engineering graduates in design, and best-practice companies provide extensive in-house training and have introduced into practice some advanced methods. The experience of industrial firms should be used to help universities improve design education and research. For example, firms could:

  • encourage universities to increase the supply of qualified graduates who are familiar with contemporary design concepts and methodologies;
  • aid in setting goals and planning curricula;
  • familiarize faculty with industrial design best practices, best processes, and the content of industry training courses;
  • encourage senior design engineers to teach in universities;
  • provide internships for faculty and graduate students; and
  • increase support of design-oriented research, including industrial participation in that research.

Support of faculty and graduate student internships should be specifically structured to provide experience in a firm's design activities. Programs that involve engineering faculty one day a week on an industrial design team are valuable, but full-time industrial experience should be encouraged for all engineering faculty members, as only full-time industry employment is likely to instill sufficient awareness of the multitude of factors that influence engineering design. Such experience has been denigrated by faculty reward practices. To take what amounts to a risky and inconvenient avenue, faculty in research universities need incentive, such as coupling industrial experience with assurance of design-oriented research support from industry or government for several years after returning to academe. Such support would significantly reduce career risk to faculty and encourage administrations to look more favorably on industrial internships. Such programs should help faculty to arrive at a broad understanding of design and manufacturing practice and lead to the identification of new research problems.[59] Faculty internships with advanced companies in Japan, Germany, and other nations should also be encouraged. The general success of cooperative education programs in engineering suggests that expansion of industrial internships for graduate students would also pay off.

Participation of designers from industry in academic work must also be expanded. One mechanism for achieving this is to fill distinguished design engineer positions on the faculty with senior designers from industry who are knowledgeable about current design best practices. Experience has shown that institutions that emphasize the metric of research funding per faculty member are reluctant to take this step, but research funding targeted at the distinguished design engineers could serve to alleviate this problem. Participation by engineers from leading foreign companies should also be encouraged.

Other mechanisms for increasing university-industry interaction in engineering design should also be explored. Joint industry-university advisory boards appointed for the specific purpose of improving design education can foster collaboration and facilitate industry assistance of university programs. As mentioned above, they can also effectively support on-campus change agents. Companies should encourage design engineers to participate on such boards.

In the long run, successful university-industry interaction should affect universities' reward systems. University-industry interaction in design should come to be viewed as an asset rather than an obstacle that prevents faculty from producing scholarly work. Moreover, industrial experience should inspire faculty to recognize the intellectual challenge of design and to generalize from domain-specific design methods to the benefit of a wide range of companies.

Summary

Design education is clearly weak; it must receive increased emphasis and introduce modern practices if it is to educate engineers who will contribute to the drive toward greater industrial competitiveness. Design issues must receive attention throughout the curriculum, and faculty must be encouraged to embrace design teaching and research. Although cooperation and additional resources will be required from both government and industry, it is important to emphasize that, particularly in the near term, educational institutions can make significant progress without waiting for additional resources.

4
A National Engineering Design
Research Agenda

Research is a central ingredient in repairing the national infrastructure in engineering design. It will contribute new knowledge, new ideas, and new people to industry and education and stimulate the creation of new business enterprises. Over time, a well-conceived, sustained program of engineering design research will gradually reduce U.S. companies' reliance on ad hoc design methods and improve their ability to produce higher quality, lower cost products and reduce lead time to market for new or modified products. It must be emphasized that the research must deal not only with the product functional quality aspects of design but also with product cost and time to market.

Research is generally divided into (1) basic research, which creates new knowledge and methods that explain or describe (often formally) natural phenomena or human behavior, and (2) applied research, which extends basic research with an emphasis on producing results directly useful to practitioners. These poles are at the ends of a continuous spectrum of research activity that ranges from the most basic, in which the goal is fundamental knowledge and understanding, to the most applied, in which the objective is to put knowledge to specific, immediate use. This chapter describes a broad topical research agenda for engineering design that ranges from fundamental research to applied topics with broad applicability. Highly directed applied research, usually driven by a specific problem, must remain the responsibility of individual firms.

THE NEED FOR BASIC RESEARCH IN ENGINEERING DESIGN

Though other sections of this report stress the need for industry and engineering education to acquire, use, and teach existing advanced design methods, much remains to be learned about engineering design processes and the knowledge and strategies needed to perform quality design quickly. Basic research is needed to generate ideas and foundations for new methods, processes, and supporting tools, and to foster continued improvement of practice. As mentioned earlier, competitiveness demands the continual development of new design methods. Using methods developed by a firm's competitors always relegates that firm to a trailing position with regard to quality, cost, and time to market of new and improved products.

Engineering design today is based largely on rather specific ad hoc bits or kernels of knowledge gathered from experience (i.e., heuristics). Though engineering design is clearly a knowledge-based intellectual activity, and its basis in knowledge and strategies should be amenable to acquisition, generation, organization, testing, and evaluation, a foundation of general knowledge, principles, and strategies has not yet been developed.[60] Once organized and generalized, design knowledge and strategy could be taught, learned, and used more effectively, and gaps in knowledge might be revealed to guide further research. Other roles for basic research include evaluating existing knowledge and strategies and providing formal principles and foundations for new tools and methodologies.

One category of basic research in the agenda presented below consists of studies that investigate the scientific foundations of design. Since skepticism about the possibility of discovering design theories, and about their potential usefulness should they be discovered, still exists both outside and within the engineering design community, it is important that the nature of research into the scientific foundations of design be addressed.

Theories, in any field, are testable, inductively generated statements about relationships among operationally defined variables or abstractions. In the physical and natural sciences, this definition is readily interpretable and well understood. Design, however, is not a physical or natural phenomenon but a complex intellectual and social process that involves many poorly understood variables, abstractions, and possible relationships. This complexity makes difficult, but does not prevent, the formulation of useful theoretical foundations, at least for important aspects of the design process.

In connection with this complexity, the field of engineering design can be viewed as consisting of three independent categories of variables and abstractions: (1) a wide variety of problem types, (2) a wide variety of persons who may be required to solve the problems, and (3) a wide variety of organizations and environments (including tools and available time) in which the persons may be required to function. Attempts to discover crucial variables and abstractions that apply to persons and the environment are likely initially to be either unmanageably complex or else greatly oversimplified. Moreover, research methodology in these categories is cumbersome and difficult to plan and implement. Obstacles faced in the cognitive, social, and environmental aspects of design are much the same as those faced by researchers in such fields as education, sociology, and management. Such design research, having been extremely limited in scope by virtue of very limited funding, has appeared "soft" and quantitatively inconclusive. Even so, it has yielded or confirmed qualitatively useful insights (e.g., most designers tend not to explore alternatives well enough).[61] Further insights that might aid organizations in planning and managing the human aspects of the design process can be expected as research methods are refined and goals are advanced. For example, principles by which concurrent design teams should be organized might be generated or the nature of the tools most helpful to human designers might be discovered.

A less subjective approach to the development of theoretical foundations for design focuses only on engineering aspects of design, omitting human and social environmental issues. Such an approach might begin with well-defined problem types and the associated knowledge and processes required to solve them. Design problems have been categorized as parametric, configuration, and conceptual. A further subdivision can be made on the basis of whether the product to be designed is a single component or an assembly of components and/or subassemblies.[62] Once a complete taxonomy of design problems is developed, and means for identifying and formulating each type are known, the search for engineering design methodologies becomes one of finding and explaining generic processes, strategies, and knowledge applicable to each type. Such problem-type-oriented fundamental development has begun and is in fact quite extensively developed for the parametric design of components.[63]

Work on parametric design of assemblies is progressing, since various optimization approaches are potentially applicable. At the configuration and conceptual design levels, some theoretical development of synthesis and evaluation processes has begun, but nothing so well established or formal as optimization theory or statistical methods has yet evolved for these problem types.

A set of viable, proven, formal processes for solving all types of design problems will constitute a theoretical foundation for engineering design that will evolve as research discovers new or improved methods.[64] Though initial progress has been made in recent years, much of it supported by the NSF's Program in Design Theory and Methodology and design-oriented Engineering Research Centers, much more remains to be done.

A TOPICAL RESEARCH AGENDA

The committee appraised a large number of potential research areas in engineering design. These areas were weighted according to potential payoff to industry and education, intellectual interest, probability of success, and required resources. Ten research topics were found to be crucially important to reforming the practice and teaching of engineering design, and thus worthy of continued expanded effort. These topics were categorized according to objective: (A) developing scientific foundations for design models and methods; (B) creating and improving design support tools; and (C) relating design to the business enterprise. Collectively, they comprise a national research agenda that will serve to guide the NSF, other government agencies, private foundations, industries, and individual researchers in the selection of research priorities and emphasis. That agenda is outlined below.

A. Developing scientific foundations for design models and methods
    1. Computer representations of in-progress designs
    2. Generating, organizing, and generalizing design knowledge
    3. Synthesis: parametric, configuration, and conceptual design
    4. Tolerance synthesis
B. Creating and improving design support tools
    1. Designer-oriented computational prototyping, analysis, and simulation tools
    2. Rapid physical prototyping
    3. Design for 'X'
C. Relating design to the business enterprise
    1. Quality-cost models
    2. Organization and communication models
    3. Innovation

Each of these 10 research areas is described in more detail below. They are not further prioritized because their value will depend upon the quality of the research performed, and their usefulness will vary from industry to industry.

A. Developing Scientific Foundations for Design Models and Methods

Research topics in this category deal with the fundamental scientific foundations on which the subsequent development of new design practices and tools will be based. Current design practice and the foundations for many existing design tools have evolved from collections of ad hoc practices and heuristics that are believed to have worked in the past or in other circumstances. Moreover, the knowledge on which most design is based is largely fragmented and unorganized and often highly specific to companies, product types, or technical domains. Formal foundations for new design-oriented CAD and solid modeling systems are needed, and fundamental studies of design models that uncover the knowledge and strategies needed to perform design will serve as the basis for improved design methodologies in the future.

A.1. Computer Representations of In-Progress Designs

A representation is a description of a design. Descriptions change during the design process from highly abstract to highly detailed. To support new best-practice product realization processes, especially at early stages and in concurrent design environments, new computer representation methods are required. The need is for a formalism that supports representation of designs at the multiple levels of abstraction and detail appropriate to different stages of the design process.[65] For example, whereas the representation might focus on functional and manufacturing issues at early stages, much later a detailed specification of dimensional and manufacturing information will be needed. The representation should also support the varied activities involved in the complete design process, including, for example, many types of preliminary and detailed functional analyses and simulations, manufacturing evaluations at many stages, cost and quality estimates, marketing and sales functions, and tool and process design. Current CAD and solid modeling systems, however advanced, are not fully utilized in industry (particularly by smaller and middle-sized firms), because they do not serve these requirements.[66] Neither do current representations adequately serve preliminary design or early analysis and evaluation processes. They are not transformable into different functional partitions and do not support the different levels of abstraction or incomplete or inconsistent designs that are common in early design stages. Current systems are well founded mathematically (a definite plus), and they support detailed analyses (e.g., finite element methods) reasonably well (though designer interfaces are still awkward). The new representation methods needed to support new product realization processes will owe a great deal to current systems and build on the knowledge and experience gained in their development.[67]

It is widely believed that the new representation methods will involve feature extraction from existing systems, and work on designing with features has begun, though the effort is small as yet. A formal definition for the term "feature," though still wanting, is expected to generalize the concept well beyond the original notion of "form features," i.e., surface elements such as holes, bosses, and fillets. The required definition should probably include information about materials, relationships to other forms, and manufacturing, as well as about form.[68]

Development of a foundation for new representations that support concurrent design, design for "X," and other aspects of best-practice product realization processes, is an extremely high-priority research need. A new generation of more "intelligent" CAD and solid modeling systems that provide designers with convenient manufacturing information and powerful, easy-to-use analytical tools simply cannot be built until research resolves the underlying representation issues.[69,70]

A.2. Generating, Organizing, and Generalizing Design Knowledge

Engineering design is a knowledge-based, knowledge-intensive intellectual activity. Designers and others involved in the design of any product or process bring to bear extensive technical knowledge, product knowledge, manufacturing process knowledge, design process knowledge, memories of previous projects, and so forth. Much of this knowledge is presently ad hoc and heuristic, residing implicitly with individuals or within organizations and neither accessible to, nor of a form that is easily accessible by, others within the firm, much less in other firms or disciplines. The handbooks, textbooks, catalogs, trade journals, research journals, and company guidelines in which much of this knowledge has been recorded are generally useful only if close at hand (some say "within reach") and if they deal specifically with the designer's current problem.[70] As a data base, this collection is extremely inefficient in terms of accessibility.

A design knowledge base more generally and completely accessible to all engineering designers would be tremendously powerful. For this vision to be realized, existing knowledge must be organized and, where possible, generalized. Once this is done, the knowledge might be made available to designers via CAD systems or computer networks. With the existing knowledge organized, identifiable gaps will serve to guide future research.

A few very small steps have been taken toward improving the design knowledge base. Some knowledge-based expert systems have been developed for specific applications, and some computer-based catalogs and design libraries are becoming available. In the future, these might be incorporated into CAD systems. There are some difficulties (i.e., research opportunities) involved in achieving the desired results here. The volume of information is huge; taxonomies of design knowledge that might serve as organizing principles for the knowledge are still lacking;[71] all the problems of very large data bases are relevant; and there will be problems with some firms' unwillingness to share information considered proprietary. Nevertheless, if the world's best design knowledge can be acquired, organized, generalized, codified, and made available to designers in a convenient fashion, design practices will not only be improved, but also made more efficient. "Reinventing the wheel" can become a phenomenon of the past except, no doubt, where direct competition prevents information sharing. Engineers will be better able to explore alternatives; educators will have a much more teachable knowledge base; and engineers will have the accessible sources of information that are essential for speedy, reliable design practices. With the design knowledge base better organized, new practices and tools based on proven models and methodologies could be continually developed.

A.3. Synthesis: Parametric, Configuration, and Conceptual Design

The object of design processes is to synthesize solutions, that is, to combine separate ideas and information into a unified whole. This process of adding and integrating information and knowledge about the design (including its function, shape, size, materials, manufacturing, and so forth) is done almost continuously, from the early, highly abstract and incomplete stages to the later, much more complete and detailed stages. Synthesis at every stage involves generating alternative solutions to the problems at hand, analyzing and evaluating those alternatives, choosing among them, and integrating the information derived into the design so that the design process can proceed to the next step. Because synthesis is so pervasive in design, it is important to understand it on as fundamental a level as possible.

Work to date has generally taken the form of developing models of design processes at various stages and/or for different domains (e.g., linkages, power plants, building structures, and so forth).[72] The stages usually studied are those mentioned earlier: conceptual design (sometimes referred to as preliminary or embodiment design); configuration design (wherein the basic arrangement of the parts of the design is settled); and parametric or detailed design (wherein the specific values for the different attributes or parameters of the design are determined). By far the most work to date has been done on synthesis at the parametric level. The field of optimization,[73] which applies here, is well developed, but its techniques are not always relevant to realistic design situations; continued work to correct this is needed. Also, optimization methods are not yet available for assemblies of parts that have important interactions or crucial evaluation issues that occur only at the system level (e.g., natural frequency).[74] Taguchi and other statistical methods for achieving robustness are also parametric design synthesis procedures.[75] Finally, a number of knowledge-based computer methods for parametric design have been developed with varying degrees of generality and usefulness.

Though most of this work has aimed at developing synthesis models and methods at the parametric stage, a complete science of parametric design has yet to be articulated. Important gaps exist that can be closed by research. At the configuration and conceptual levels, very little has been done even to develop synthesis models and methods. Physical principles, at least qualitatively, are involved in both conceptual and configuration design. To date, the little work that has been done has been limited to narrow, domain-specific studies.

At every stage, synthesis involves the generation of alternative solutions, that is, innovation (discussed in Section C.3), evaluation, and decision making. These basic processes as they apply to engineering design need more study and integration into design process models and methods. New synthesis models and methods for various types of design processes can lead to new and improved best practices. We must continue to refine methods for the parametric stage and greatly increase the study of all aspects of synthesis (e.g., innovation, decision making, evaluation methods, knowledge and strategies needed, and so forth) in earlier stages of design.

A.4. Tolerance Synthesis

Although it is an aspect of parametric design, tolerance specification is such a crucial driving factor in product cost and performance that it deserves special attention.[76] Tolerances are applied to nominal dimensions of a part or product to indicate allowed divergence from a nominal value. Designers need readily usable information and procedures that support the judicious assignment of tolerances to optimize tradeoffs between product performance and cost. Though tolerances can now be set rigorously in a few highly specific cases, most tolerances are based on experience and company customs that reflect a mostly subjective attempt to balance product performance and manufacturing cost. There often exists a great deal of company-specific data on tolerance-cost relationships, but little general data and even less data that relate tolerances to product performance. Engineering designers are thus usually in the position of assigning tolerances without the benefit of solid data or rigorous theory. Some initial research, if continued, might yield the ability to represent tolerances appropriately in CAD and solid modelers, though this is a difficult task. Deficiencies in current tolerancing methods have been revealed in the process of applying them to mathematically rigorous solid models. Tolerance analysis, an essential aspect of tolerance synthesis and an extremely complex process, especially in three dimensions, is not yet fully developed. Finally, tolerance standards are not always consistent with available and evolving measurement methods.

Research is needed in tolerance analysis, tolerance representations, tolerance-cost relationships, tolerance-performance relationships, and tolerance standards and measurement methods. On the foundations laid by this research it will be possible to build design support tools to aid designers in making optimal tolerance selection decisions.

B. Creating and Improving Design Support Tools

The introduction of new tools that improve designer productivity or performance is the most direct cause of changes in design practice. Both the research in the previous section (which will provide foundations for new tools) and the research described here (to result more directly in new tools) are needed to create new tools. There is only a blurred and sometimes arbitrary distinction between research into foundations for new tools and the development or improvement of the tools themselves; this section of the research agenda comprises those subjects closer to tool development.

B.1. Designer-Oriented Computational Prototyping, Analysis, and Simulation Tools

Analysis and simulation are supporting elements of design processes; they provide data and information about behavior, functional performance, cost, manufacturing, and other issues that are essential to intelligent design decisions. Although many computer-based analysis and simulation methods are available, especially for the detailed stage of design, few are in widespread use, particularly in smaller and middle-sized firms. One reason is that the technologies that employ these methods are not workable in all computer environments. Another is that proper use of these methods and tools requires highly specialized knowledge. Most analysis and simulation methods and tools have been developed for use in the final detailed stage of design. There is a strong need for these in many situations, but as emphasis on decision making shifts to earlier stages (as in concurrent engineering), there is an equally great need to provide new analysis and simulation methods and tools that are applicable before a design is completely specified.[77] Research is needed to develop such new methods and make them available to designers. Computational prototyping (i.e., the ability to experiment with the behavior of parts or products using their computer representations) reduces design cycle time by reducing the need for actual physical prototyping. New representations, as discussed in Section A.1, may be developed that will support powerful computational prototyping tools, including on-line handbooks and catalogs to increase efficiency further.

Here, the needed research is (1) to develop new methods of analysis, simulation, and computational prototyping that serve early stages of design and the new concurrent design practices, and (2) to make both existing and new tools readily useful to all designers.

B.2. Rapid Physical Prototyping

Although analysis, simulation, and computational prototyping aim to shorten design and product development cycles and to improve the quality of the results by doing as much product testing as possible on the computer, ultimately a physical prototype must often be fabricated. Thus, tools are needed that link design and manufacturing quickly and inexpensively for prototype construction. Prototypes can serve different purposes at different stages of the design process. One may serve to test the applicability of a new material or process; another may test tolerance issues; some will have multiple purposes. Physical prototyping methods and tools are needed that serve the specific needs of the design process and that enable rapid realization of the desired physical model. The MICON system at Carnegie Mellon University exemplifies the state of the art for electronic systems.[78] An analogous system from the mechanical domain is Kimura's variant process planning system developed at the University of Tokyo.[79] First-Cut, a system under development at Stanford University,[80] bridges the gap between CAD and CAM by supporting simultaneous design of a product and the process used to manufacture it. Commercially available stereolithography techniques can produce complex parts from CAD representations, though material issues are not readily evaluated by this means.

Research is needed to define the various types of prototypes and their purposes, and practical, low-cost, and rapid methods must be developed to meet the needs of each type. For example, for processes such as injection molding, disposable dies may be feasible. Means of reducing the need for physical prototyping can be explored. The goal is to develop methods and tools that enable firms to construct physical prototypes, when necessary, both quickly and inexpensively.

B.3. Design For 'X'

A product must satisfy many objectives: function as perceived by the consumer; ease of assembly; maintainability; testability; safety; disposability; and many others. These are the X's in "design for X." In best design practice, all are considered at the earliest stages of design as well as continuously throughout the design process.[81]

The first X to receive explicit attention was assembly. Design for assembly (DFA) methods and tools developed by Boothroyd are widely (though not yet fully) disseminated.[82] Methods and tools to support design for manufacturing (DFM) in processes such as injection molding and forging have also now been developed. These methods identify, through experimentation, experience, or insight, the crucial features that affect ease or cost of manufacture of parts and assemblies, and then the presence, configuration, or parameters of these critical features are related to assembly or manufacturing time or cost.

This approach can be extended to many more X's and to the entire design and product development process.[83] That is, the critical features that influence each X at each stage of design need to be identified and related specifically to their impact on X throughout the life cycle of the product. At this time, DFA and DFM knowledge relates primarily to the parametric stage of design, though some aspects apply to the configuration stage. Designing for manufacturing at the conceptual stages and designing for other objectives (i.e., X's) at almost any other stage are not supported by any well-developed techniques. Consequently, consideration of these objectives tends to occur only after major design commitments have been made. Studies are needed that seek to relate the crucial features of a product's early description to its ultimate life cycle quality and cost in terms of each of the many design objectives (X's).

C. Relating Design to the Business Enterprise

Research on design in a business context addresses issues related to understanding and supporting design and product development in a companywide context, thus recognizing that functional and manufacturing aspects of a product cannot be considered independently of personnel, marketing, finance, accounting, and other business issues.

C.1 Quality-Cost Models

Quality and cost models are coarse-grained, but realistic models of relationships between manufacturing costs, time to market, user costs, and quality aspects of a product or process. Useful models are particularly needed at the earliest stages of design to support tradeoff studies and management and engineering decisions.[84] Accurate accounting for indirect costs and internal transfer costs is also important in these models.[85] At present, very few cost models and even fewer quality models are available that capture key cost, quality, and time drivers and their relationships at the conceptual design stage. Cost models are available for later stages, but these do not aid early design decision making. Quality models are generally lacking for all stages. Quality-function deployment,[86] used by some firms at the conceptual stage, is currently highly subjective but could be made more accurate and more widely applicable through research. Taguchi's quality-loss function is used by some firms at the parametric stage. Traditional cost accounting based on unit labor and material costs is usually unrealistic and can lead to inappropriate design and product development decisions. More research, possibly along the lines of activity-based management accounting systems, is needed.

Development of more accurate, tested quality and cost models, especially at the preliminary or conceptual design stages, is essential to support effective concurrent design. Studies are needed that identify early the product features that drive downstream quality and cost and relate these features to ultimate quality and cost, allowing models to be developed that support tradeoff decisions throughout the design process.

C.2 Organization and Communication Models

Organizational issues in design relate to the planning, organization, and management of product realization processes, including the creation and use of cross-functional or interdisciplinary teams. Communication issues relate to the facilitation and control of information transfer, both internally and externally, in design projects.[87]

Several decades of work in organizational studies conducted by departments of psychology and sociology in universities and by organizational groups in management schools have seen little focus on design or even on engineering. Such work that has dealt with design has produced some useful results in the form of prescriptions for the organization of technical projects, design of facilities, structure of information systems, and organization and management of teams, but current work is weakly focused. Very little has been done, for example, on the flow of information within a design organization or between a manufacturing firm and its suppliers, yet these and other organization and communication studies are relevant to other areas of design research and to the development of better computer-based supporting tools. The goals of research in this area are several: to create and evaluate useful models of how information is and should be exchanged and used in a product realization process; to understand how multidisciplinary teams work in order to improve their performance; to create and evaluate models of product realization processes; and to learn how various supporting tools influence the performance of teams and of a product realization organization. The payoff will be shorter design cycles through improved organization and communication effectiveness. Information about cross-disciplinary teams can be used to support the design education process.

C.3. Innovation

Innovation is the generation and implementation of new, unique solutions to stated problems. It is generally agreed that innovative capability is valuable in product development processes, but there is little agreement on how to stimulate it or on how to evaluate its cost-benefit tradeoffs.

Studies of the innovation process in individuals conducted in a variety of domains have yielded techniques for fostering innovative individuals, but little is known about the effect of innovative individuals on teams and vice versa. There is also little understanding of how to organize and manage groups and design processes so as to encourage innovation; factors that influence innovation in teams, such as team composition, time allotments, group environments, reward policies, and so forth, have not been established. It is conventional wisdom that most industry situations allow too little time to be spent on innovation, but data to support allocation of more time are not available.

It is likely that knowledge is a key ingredient in innovation, both at the individual and group levels, but this too has been little studied.

Research in innovation is difficult, especially if it is to be credible in and relevant to industrial settings. Studies must be carefully planned and should be directed at establishing the value of applying resources to stimulate innovation and developing and evaluating methods for increasing innovation in groups and organizations.

Benefits of Implementing the Engineering Design Research Agenda

The benefits to design practice and education that can flow from implementation of this research agenda include:

- a new generation of computer-aided design tools that support preliminary as well as detailed (parametric) design and that provide designers with information needed for manufacturability and life cycle issues;
- prescriptions to improve organization and communication in the product development process;
- useful quality-cost models that can support design and management decision making at the preliminary design stage;
- improved interfaces to engineering analysis and simulation tools;
- better information relating tolerances to cost and performance;
- more complete and implementable methodologies for design problem solving;
- new prescriptions for generating and evaluating configuration and conceptual design alternatives;
- greater availability and accessibility of the knowledge needed by designers to perform all types of design;
- development of design procedures that lead toward integration of all stages of design, from concept through disposal, involving the entire business enterprise, and directly addressing concerns regarding cost, quality, and length of the design cycle.

The importance of the proposed research on these 10 areas to the revitalization of the engineering design infrastructure in the United States and hence to U.S. competitiveness cannot be overemphasized.

Resources Required

Though most of the recommended research is open-ended and should continue for many years, significant and useful intermediate-term (i.e., four to five years) results should be achievable in almost all areas. The best assignment of resources is to supply for each topic six to nine groups of researchers each consisting of two to four professionals. On average, funding per topic comes to about $2 million annually for four or five years. Total project commitment is thus $20 million annually for four or five years and between 120 and 360 researchers. Because a sufficient number of researchers may not be available, it may not be possible to begin all research immediately.

It is extremely important that all of this research, whether applied or basic, be of the highest quality. Researchers and sponsors must ensure that important research issues and problems are defined and rigorous research methods are followed. To make these research efforts truly responsive to industry needs and to familiarize university researchers with the connections among design and manufacturing, vendors, customers, sales, and service, the research should involve frequent and close interaction between researchers and design engineers in industry.

Both industry and academic communities will have to be willing to reach out and engage in the communication necessary to achieve mutual respect and understanding. Industry representatives will need to value and appreciate, become involved in, and provide support for intermediate- and longer-term research efforts. Engineers in industry need to read and contribute to the research literature related to design. The academic community needs to appreciate that design in industry takes place in the context of highly competitive business enterprises, a fact that has important implications for research. Finally, research results must be disseminated with industrial as well as academic readers in mind.

DISSEMINATION OF RESEARCH RESULTS TO INDUSTRY

Engineering design research can yield major advances in engineering design methods, but the research must be related to the problems of industry and must be readily adaptable to the industrial design environment. Though university research efforts in engineering design are frequently long range and their results potentially applicable across a wide spectrum of industries, problems with dissemination of research results have left most engineering designers and engineering design managers believing that current design research has little relevance, so they are unwilling to seek and utilize new research results.

There are a number of paths by which the results of university research in engineering design might be brought into industrial practice. One route is through the development of new engineering design support methods and tools based on the fundamental knowledge generated by basic research.[88] This process is often slow, however, and can require intensive development work beyond the abilities of university research laboratories. A second route is through new M.S. and Ph.D. graduates whose education, including their thesis research, has stressed engineering design. These graduates can either enter industry, bringing with them new and advanced engineering design knowledge, or accept faculty positions, through which they can pass their knowledge to a new generation of practicing engineers.[89] A third route is through design-oriented faculty members who work or consult in industry and engineering designers from industry who spend time in universities.

Although these personal modes of information transfer are important and contribute to awareness of new ideas in industry, effective exploitation of engineering design research demands that research results be put into forms useful to industrial firms. The new methods must be refined and packaged as products (mathematical or statistical computer packages, CAD systems, expert systems, and so on), a task not readily performed either by most universities or by most of the companies that might use the research. A few (generally small) firms have developed some research results into tools and methods usable by industry; other entities that might perform this development activity include government-funded organizations, multiple private design-oriented companies,[90] industry consortia, start-up firms, or some combination of these. The creation or enhancement of such research transfer paths would especially help small and medium-size companies that are unable to perform the task of development internally.

A mechanism is also needed to perform applied design research, sometimes referred to as precompetitive research, and disseminate the results.[91] This type of research is directed at industrywide problems that are too large or "too applied" for university laboratories and not amenable to cost-effective resolution by a single company.

Without some sort of organization acting as broker, the results of research on engineering design are not likely to reach the greatest number of potential users. An organization is needed to gather and disseminate information about international best engineering design practice, perform research to improve design methods and tools, and promote design technology transfer. This organization might also help arrange personnel exchanges and arrange privately funded research between universities and industry.

A NATIONAL CONSORTIUM FOR ENGINEERING DESIGN

The committee discussed many ideas, methods, and techniques for dealing with the dissemination of research to industry, conducting precompetitive research, and acquiring and disseminating the world's best design practices, as well as the need for greater interaction between universities and industry and for brokering agencies to encourage such interaction. A National Consortium for Engineering Design (NCED) was considered for dealing with these issues. The following list characterizes, but does not limit, the potential charter of an NCED.

1. Acquire or develop the world's best engineering design practices and processes.

2. Acquire or develop the world's best computer-aided engineering, design, test, materials handling, manufacturing, program planning, and other tools.

3. Document processes, practices, and tools, create training materials, and provide training at several levels (e.g., for trainers, practitioners, and managers).

4. Develop research proof-of-concept software into robust, user-friendly software ready to be used in practice or to be commercially developed for transfer to industry.

5. Conduct engineering design research as described in the research agenda.

6. Provide "hands-on" opportunities to learn by executing new designs.

7. Provide expert support and problem solving capabilities to members.

8. Facilitate collaborative corporate, government, academic, and NCED projects.

9. Establish *industry-led* engineering design applications projects that provide university faculty, graduate students, and government employees with industrial design experience.

10. Establish *university-led* engineering design research projects that provide industrial and government people with research experience directed at creating new design practices and tools.

11. Establish *government-led* engineering design projects to provide industrial and university people with experience in government sourcing processes.

12. Provide on-site courses taught by university, industry, and government people, as appropriate (e.g., a graduate course that can call on government and industry people, as well as professors, to lead lectures and workshops).

13. Host a yearly conference of engineering deans, industrial chief engineers, and human resource directors and government research agencies at which university course content and research directions, industry design and education practices, and personnel exchanges are discussed.

The NCED would be a nonprofit organization, funded by participating industrial, government, and academic organizations. It would, when fully operational, have a board of directors/trustees drawn from industry, academe, and government, as well as from the NCED itself, full-time administrative and technical staffs, and full and part-time representatives from participating organizations. NCED would study and perform research on mechanical structures, opto-electro-mechanical systems, and some widely used materials, and it would have enough manufacturing capacity to support concurrent design efforts. The output of NCED would be delivered through documentation, training, consulting, expert participation in member development programs, member visiting appointments to NCED, graduate student programs, sabbaticals, and other appropriate modes.

The NCED environment would lie somewhere between that of a major research center and that of a product development organization. A higher level of support and teamwork would be expected than is traditional within universities or corporate and government research laboratories. Joint projects, carried out under the leadership and at the site of either the member enterprise(s) or NCED, would be encouraged. Intellectual property rights would be negotiated in favor of the sponsoring members. Incrementally funded proprietary projects would be facilitated, and NCED employees would be encouraged to consult for some portion of each month. The primary objectives of the NCED would be to develop and accumulate knowledge of world-class engineering design practice and processes and transfer that knowledge to sponsoring organizations through a variety of formal and informal mechanisms.

5

Recommendations

Poor engineering design capabilities are leading U.S. companies to design and produce products that are more expensive, of lower quality, and slower to reach the market than those of their top foreign competitors. This situation can be corrected, but the task will not be simple. Universities, with few exceptions, are no longer preparing graduates adequately in design. Too little engineering design research is being conducted, it is often dissociated from industrial needs, and its results are poorly disseminated. Few companies have coherent product realization processes, implement available advanced design techniques, or develop new ones.

Greatly improved engineering design is so important to the nation's industrial competitiveness and economic health that government must proactively emphasize it as part of a long-range emphasis on commercial excellence and national competitiveness. General advancement of engineering design practice does not favor one firm or one industry over another. It benefits the entire society, including individual consumers, even in areas where foreign competition is not yet a factor.

Though there is no one logical funding source for engineering design, NSF, the Department of Commerce, the Defense Advanced Research Projects Agency, and industry all have roles to play within their missions. Until there is a broad realization in government that sectors of the nation's civilian technology base are in jeopardy, and that support of engineering design is a high-leverage area where government support will produce large benefits, these agencies will have to take the responsibility of coordinating one another's piecemeal efforts to improve engineering design.

The following recommendations are intended to set in motion a number of separate remedial activities in industry, academe, and government that will improve industrial engineering design practices to increase competitiveness and create new leaders in design. Some of these recommendations will have an immediate effect when implemented, but it will take some time to rebuild the design infrastructure. It is thus imperative that action be taken now.

IMPROVING DESIGN PRACTICE

Though all the recommendations presented in this report will eventually aid industry, companies can take a number of steps to improve their own engineering design capabilities, and thus their competitiveness.

As discussed in Chapter 2, *manufacturing firms* should:

• recognize the leverage afforded by engineering design and move to take advantage of it;
• implement a comprehensive, coherent product realization process;
• utilize a carefully chosen set of contemporary design practices;
• create a supportive environment for design;
• establish dedicated functional change agents to implement new practices and organizations;
• actively promote and support continuing education of their engineers;
• aggressively support research and development activities in engineering design;
• continually and formally seek and incorporate the best practices as they evolve; and
• adopt modern management accounting systems.

IMPROVING ENGINEERING EDUCATION

Chapter 3 emphasized that engineering design education in the United States is poor and that strengthening engineering design education is critical to the long-term development of engineers who are prepared to become good designers and leaders who will provide a lasting foundation for U.S. industry's international competitiveness. The recommendations that follow deal with improving both engineering curricula and the teaching of design.

Curricula

Engineering institutions in the United States must improve the design component of engineering curricula. The following reforms are targeted at undergraduate programs, but many also apply to graduate programs. Each institution should:

• ensure that its engineering curricula fully meet both the letter and the spirit of current ABET accreditation criteria related to design;

- establish evaluation metrics for the design component of engineering programs and mechanisms to monitor performance in terms of these metrics;
- ensure that design courses cover best international design methods;
- utilize carefully appointed external advisory boards of engineers and engineering managers from best-practice companies to monitor and improve the design component of curricula;
- ensure that students are prepared to continue learning throughout their careers;
- create collegewide change agents to spearhead the efforts to improve design education and assist in gathering and developing instructional materials, promoting interdepartmental and university-industry cooperation related to design, and establishing metrics for evaluating design in various curricula; and
- place in faculty positions senior engineers from industry who are knowledgeable about current and evolving design best practices.

Industrial firms need to raise university awareness of industry needs in engineering design by:

- making clear to universities that they want graduates who are familiar with contemporary design concepts, principles, and methodologies;
- explaining to universities their best design practices, what they believe should be taught, and what they are currently teaching their own engineers;
- encouraging their designers to participate on university advisory boards and teach in the classrooms; and
- supporting design education by providing opportunities for faculty and students to observe and participate in design experiences, both in industry and academe.

ABET and the engineering societies that are ABET participating bodies should:

- stimulate the further incorporation of design into engineering curricula by changing its criteria for accrediting programs in engineering in Section IV.C.2.d.(3), pertaining to the minimum one-half year of engineering design, as follows (recommended deletions are shown by strikeouts and additions are italicized):

"The engineering design component of a curriculum must include at least some *nearly all* of the following features: development of student creativity, use of open-ended problems, development and use of design methodology, formulation of design problem requirements and specifications, *in-depth* consideration of alternative solutions, feasibility considerations, *production processes, advanced design methodologies, concurrent engineering design, life cycle considerations,* detailed system descriptions, *and participation in an interdiscipli-*

nary group on a design project. Further, it is essential to include a variety of realistic constraints, such as economic factors, safety, reliability, aesthetics, ethics, and social impact. *Finally, design courses should be integrated into the curriculum so as to provide a continual increase in the design competence of a student progressing through an engineering program."*

Professional engineering societies should:

• through their education arms and with participation of engineers practicing in industry, encourage the further education of design teachers and increase the awareness of all faculty members of the importance of engineering design.

Support for Faculty

Engineering institutions must adopt organizational changes that facilitate and reward design teaching and research. They must:

• modify reward systems so that they recognize the contributions of teachers and researchers in engineering design;
• remove impediments to interdisciplinary and interdepartmental collaboration in design education; and
• encourage faculty to participate in design education training and workshops in both industry and academe.

The National Science Foundation should facilitate improved teaching of design by establishing a clearinghouse for design instructional materials and methods. The mission of the clearinghouse would be to:

• collect information on best design practices and research worldwide;
• facilitate the synthesis of this material into textbooks and problem sets, case studies, descriptions of modern design theory and practice, video tapes, computer software, course outlines, and candidate curricula;
• publish reviews of design research, teaching methods, and software tools;
• facilitate the introduction of standards and common representations (e.g., IGES, PDES); and
• actively disseminate the results of all the above activities through all appropriate means.

Implementation of this clearinghouse should proceed quickly, possibly as an adjunct to some existing design program. If the consortium (NCED) discussed later in this chapter becomes operational, the clearinghouse might be incorporated into it. The information disseminated by this clearinghouse could be valuable for industrial practice and research, but it must be cast in a form appropriate for faculty teaching use.

IMPROVING ENGINEERING DESIGN RESEARCH

Research is a central ingredient in repairing the national infrastructure in engineering design. Because it has been largely neglected for decades, a strong, high-quality research initiative is especially critical at this time.

Aggressively Pursuing the Research Agenda

The *National Science Foundation* should propose, and *Congress* should fund, an Initiative for Engineering Design to support both a large increase in design research and increased university-industry interaction in engineering design. The major thrust of this initiative should be an expansion of university research in engineering design. One facet of the initiative should be support for a Design Scholar program that would enable university faculty and Ph.D. students to spend one to two years with a best-practice industrial firm, followed by three years of NSF research support with matching industry support.

Design research clearly requires such an initiative; it will be difficult for the NSF, the logical funding agency for much of the proposed research, to allocate substantial funds to support design because various NSF constituencies cannot be expected willingly to accept funding cuts in their areas, and the "proposal pressure" that drives some reallocations is not likely to be strong in an area characterized by a short history and limited past funding. For applied research, particularly research dealing with data bases, standards, and the relation of design to the enterprise, the *National Institute of Standards and Technology (NIST)* can provide much leadership and support. The *Advanced Civilian Technology Agency* that has been proposed in bills introduced into both the 100th and 101st Congresses[92] should, if created, have design research and technology transfer as one of its major activities. The *Department of Defense* and, to a lesser extent, the *Department of Energy* are supporting significant design research which should be continued and expanded.

The *National Science Foundation* should expand and emphasize its Design Theory and Methodology Program by providing a clear identity and strong leadership. Stable, continuous funding, beginning at approximately $6 to $8 million annually should be provided. The NSF program should primarily support "research on scientific foundations for design," as presented in the research agenda in Chapter 4 of this report, but basic research efforts in the other areas would also be suitable for NSF support. Interdisciplinary design research as conducted by some Engineering Research Centers, featuring widescale industrial cooperation and partnership, as well as the initiation of additional design-related Engineering Research Centers should be strongly encouraged.

Conducting Research

All *engineering design researchers* should

- be aware of the research agenda, how their research fits into it, and what the long-term goals of their research are with regard to engineering design practice;
- collaborate with industrial firm design engineers to define research topics and strategies; and
- do more to disseminate their research results, including publishing research results not only in the refereed literature but also in periodicals that are widely read in industry.

Industrial firms must take an initiative in fostering research collaboration with universities by:

- participating in, and supporting, basic as well as applied engineering design research;
- developing strategies for assuring long-term stable interactions with and support of researchers; and
- supporting faculty internships in industry, accompanied, where appropriate, by assurance of design-oriented research support from industry or government for several years after the faculty members return to academe.

National Consortium for Engineering Design

In Chapter 4, the creation of a National Consortium for Engineering Design is suggested for the purposes of:

- performing precompetitive research to improve design methods and tools;
- gathering and disseminating information about international best engineering design practices;
- transferring existing and new design knowledge, especially in the form of software, into industry and academe;
- developing and promoting industry-university-government collaboration in research and education; and
- providing brokerage services for personnel exchanges and arranging privately funded research between universities and industry.

Industrial firms have already formed several organizations for exchanging and disseminating technical information, generating knowledge of a precompetitive nature, or for other purposes similar to those outlined above. Examples of such consortia and cooperatives of various kinds include:

- The National Center for Manufacturing Science (NCMS)
- The Design Institute (United Kingdom)

- The Heat Exchange Institute
- Tubular Heat Exchanger Manufacturers Association
- National Federation for Computer-Aided Manufacturing (NFCAM)
- Integrated Program for Aerospace Vehicle Design (IPAD)
- Sematech
- Microelectronics and Computer Technology Corporation (MCC)

These cooperative groups represent a broad range of organizational forms and funding arrangements, but all were formed with the objective of helping a particular set of industrial firms to share in the generation and dissemination of technical knowledge.[93]

The NCED needs to possess an organization and operating style that allows it to be primarily industry-led and applications-driven. Among the structural possibilities are:

- an organization of industrial firms of all sizes, with funding from the firms on some proportionate basis;
- a similar organization, but with strong Department of Commerce involvement;
- an organization of industrial firms, but with government funding for start-up purposes;
- an extension of the NSF Engineering Research Center (ERC) program. More detailed study is needed to shape the organization and operation of NCED to ensure early and continuing success. Such study should be undertaken promptly.

The *Department of Commerce* and the *National Science Foundation* should, with the assistance of industrial and academic representatives, jointly study the possible structuring and operation of a National Consortium for Engineering Design for the purposes listed above.

Appendix A
Examples of Product
Realization Processes

POLAROID'S PRP

Polaroid's PRP, the Product Delivery Process, is a strategy for assuring that the essential business and technical considerations related to a product's development are considered, evaluated, and understood by the total corporation.

Each of the following elements is essential in the product realization process, and each requires attention and commitment by all levels of the company. The sequence of steps is important. It assures that the product specification be defined and agreed to before the design phase begins and that the product does not require scheduled inventions to stay on schedule.

1. Explore the business, marketing, and technical opportunities.

2. Define the customers' needs while continuously improving the product development process.

3. Define a long-range, customer-focused product line strategy and define the system's architecture for the family of future products.

4. Clearly and fully define the product performance specification with the product development team (manufacturing, marketing, engineering, finance, led by the program manager).

5. Insure that the product definition does not require inventions.

6. With clearly defined and agreed-to product specifications up front, there should be no performance specification changes during the design process (other than ones critical to customer needs).

7. Establish a benchmark process containing goals and driven by the need for continuous process improvement.

8. In parallel, continuously develop core technology building blocks for future products.

9 Design the first system layout with CAE/D/M tools from the start, utilizing multidisciplined, professionally trained engineers and designers.

10. Concurrent with the product design process, design the manufacturing process.

11. Build a reusable math model base for the product technology and use it for simulation, analysis, and modeling of future product designs.

12. Develop an information process for tracking world-class engineering design practices and share successful generic design processes with universities and other U.S. companies.

HEWLETT-PACKARD'S PRP

An important part of Hewlett-Packard's PRP is the Phase Review Process which assigns responsibilities to appropriate individuals at each stage in the development of a product. Senior managers are involved and made responsible for approving product designs. System team members can reside in different divisions and at different locations within the company. This puts the decision making in the right hands and reduces delay and contention. (AT&T uses a similar process, which it calls the checkpoint process.) The full PRP is designed to include all the important participants. For example, the designers are involved through the Break-even Metric described on page 23 and in Figure 8.

The process has a defined structure. Major management milestones and commitments are identified, including phase exit objectives and functional activities and deliverables. Signed agreements are required from approvers at each system phase exit. An escalation process is defined for issue resolution.

The process defines roles and responsibilities. System management's role focuses on company business issues and risks. Functional reviewers identify issues and commit to system readiness to exit. Senior management approvers agree upon system advancement and make corporate commitments.

The phase review process has 7 phases:

0. Requirements/Plan
1. Study/Define
2. Specify/Design
3. Develop/Test
4. User Test/Ramp Up
5. Enhance/Support
6. Maturity

They are defined and related to each other as shown in Figure 9.

Phase Review Process

	Phase 0	Phase 1	Phase 2	Phase 3	Phase 4	Phase 5	Phase 6
	Requirements/ Plan	Study/ Design	Specify/ Design	Develop/ Test	User Test/ Ramp Up	Enhance/ Support	Maturity
Key Exit Objectives	Describe requirements	Define a solution	Design solution	Develop solution	Test	Support	Replace
Key Completed Activities/Deliveries at Phase Exit	System requirements	System specifications	Functional plans	FURPS+ verification	Functional sign-off / Field preparedness	ROI verification	Migration discontinuance plans
Key Corporate Commitments at Phase Exit	Business plan	Funding a specific system	Funding functional plans	Price, performance, availability	Customer shipments, enhancement, support	Customer support	System alternatives

BUSINESS — SPECIALIZED SUPPORT — PLANNING

Figure 9: Phase Review Process Themes

Appendix B
Course Outline for
Contemporary Engineering*

The Engineer's Environment/Engineering Practice:
"What The Practicing Engineer Must Deal With In Today's Environment."

List of Topics

1. *Design*

   Philosophy of Design
   Manufacturability
   User-Friendly Presentation to Operator
   Worst Case Testing
   Reliability
   Aesthetics/Appearance/Industrial Design
   Cost vs. Pricing
   Utility
   Product Design
   Appropriateness to Market
   International Considerations
   Standards: Safety & Industry

2. *Legal*

   Product Liability
   Environmental/Pollution
   Contracts
   Ethics

---

*Preliminary draft developed by Dr. Joel Spira, Lutron Electronics Company, Coopersburg, PA. January 6, 1990.

3. *Intellectual Property*

Importance of Intellectual Property (I.P.)
Patents
Trade Secrets
Copyrights
International Trademarks
Six Significant Cases of I.P.

4. *Economics*

Keynesian
Supply Side
Inflation
Interest Rates
Business Cycles
Industry Cycles
Recession/Boom
International
Currency Rates/Consequences

5. *Marketing*

Quality, Price, Delivery
Size and Definition of Markets
Distribution
Creation of New Markets
Creation of New Goods and Services
Sales
Service
Pricing
International
Demographics

6. *Quality*

Garvin - Define
Contribution of:
    Deming
    Juran, Crosby
    Japanese
Statistical Quality Control

7. *Financial*

   What a P & L looks like/means
   What a Balance Sheet looks like/means
   Capital Formation
   Cost Accounting
   Cost Control

8. *Technology & Science* (T & S)

   Brief History of T & S since 1900
   Brief History of T & S since World War II
   Current T & S issues and consequences
   U.S. T & S vs. Europe and Asia

9. *Manufacturing*

   Relationship to:
       Quality
       Design
       Cost
       Delivery
   Flexibility to Market Forces
   Domestic vs. Offshore

10. *Impact of the Computer*

    Assembly
    Materials Control
    CAD/CAM

11. *Marketing Relation to Manufacturing*

    Flexible Specialization (Pioro-Sabel)
    Niche (short run) vs. General Market
        (high volume, low cost)

12. *New Character of the Work Force*

    (More educated and skilled and also
        less educated and skilled)

13. *New Materials and Manufacturing Techniques*

14. *Manufacturing and Purchasing is now an
    international worldwide operation*

Glossary

ABET
Accreditation Board for Engineering and Technology, the organization that accredits engineering curricula in the United States

Applied research
Extension of basic research with a focus on a perceived practical need

Basic research
Research that creates new knowledge or techniques that explain natural phenomena or human behavior, or aid in their application to human needs

Best practices
Any collection of advanced engineering design techniques that provides design excellence and high-quality products for a given product line or company

Concurrent design
Engineering design practice that combines the concerns of marketing, functional product and process design, production, field service, recycling, and disposal into one integrated procedure

Continuous improvement
A process by which products and processes are improved year after year through study, application of sophisticated techniques, and experience; applied to the product realization process, it reduces development cycle time and final cost of the product

DARPA

Defense Advanced Research Projects Agency, responsible for basic and applied research associated with the needs of the Department of Defense

DFX, Design for "X"

A collection of techniques for emphasizing aspects of design such as assembly, plastic molding, field repair, and so forth (the X's); distinct from traditional design focus on product function

Engineering analysis

The technical process by which the behavior, performance, quality, and cost of an entity are predicted on the basis of engineering descriptions and specifications (the reverse of design)

Engineering design

The technical element in the product realization process that involves the application of knowledge and techniques from engineering, science, aesthetics, economics, and psychology in establishing specifications for products and their associated production processes; the technical process by which engineering descriptions and specifications are formulated to ensure that a product will possess the desired behavior, performance, quality, and cost (the reverse of engineering analysis)

Engineering design practices

The collection of knowledge, techniques, and computer aids available to designers in pursuit of their profession; examples include concurrent design, design for assembly, Taguchi methods, quality function deployment, the six sigma method, solid modeling

Heuristics

A collection of ad hoc bits or kernels of knowledge gained from experience

Just-in-time (JIT)

A method of manufacturing by which parts and assemblies are made or delivered as needed, thereby greatly reducing inventory

Life cycle cost

The total cost to society of an item over its entire life, from initial concept through manufacturing and use to disposal

Manufacturing

The process of producing component parts, assemblies, and complete products, including fabrication, assembly, test, storage, and distribution

NIST

The National Institute of Standards and Technology, formerly the National Bureau of Standards, is part of the Department of Commerce. The change in name was put into effect under the *Omnibus Trade and Competitiveness Act of 1988*, as signed into law on August 23, 1988

NSF

The National Science Foundation, the federal agency responsible for promoting and advancing scientific and engineering progress in the United States

Precompetitive research

Research of high potential value but of such breadth and risk that the resources required to sustain it are unlikely to be available at a single location; it may be too applied for most universities and yet too remote from specific applications for a single industrial firm to support, thus requiring social (governmental) support

Product development cycle

The time it takes to create and bring to market a new product design

Product realization process

The process by which new and improved products are conceived, designed, produced, brought to market, and supported. The process includes determining customers' needs, translating these needs into engineering specifications, designing the product as well as its production and support processes, and operating those processes

Quality function deployment

A process for systematically translating customer requirements into appropriate technical requirements during all stages of product development from the earliest stages of product design through production

Quality loss function

A calculation of loss of quality as a function of deviation from desired performance; usually a continuous, not discrete, function

Six sigma method

A statistical method for quantifying the degree of deviation permitted by parts, products, and processes that guarantees that failure will typically occur less than three times in a million opportunities

Solid modeling

A technique for representing the properties of solid objects in a computer model

Taguchi methods

Generic term covering a variety of methods for statistically determining required quantitative features of a design or a manufacturing process that render it robust against disturbances, variations, and uncertainties, with the objective of reducing quality loss

Total quality management (TQM)

A set of principles having a primary purpose of increasing value to the customer and involving continued attention to quality at every step of the product realization process by all members of the organization

Bibliography

Accreditation Board for Engineering and Technology, Inc. Annual Report. 1989. *An annual publication of current criteria for accreditation, a listing of accredited programs, and an analysis of accreditation actions during the preceding year.*

Accreditation Board for Engineering and Technology, Inc. Engineering Education Answers the Challenge of the Future. Proceedings of the National Congress on Engineering Education. 1986.

Agogino, A. M., and A. S. Almgren. Symbolic computation in computer-aided optimal design. Pp. 267-284 in Expert Systems in Computer-Aided Design. J. S. Gero, ed. Amsterdam: North-Holland. 1987.

Allen, T. J. Managing the Flow of Technology. Cambridge, Mass.: MIT Press. 1977.

American Electronics Association. P. 98 in America's Future At Stake: Winning in the Global Marketplace. 1989.

American Society for Engineering Education. A National Agenda for Engineering Education. Washington, D.C. 1987.

American Society of Mechanical Engineers. Research Needs and Technological Opportunities in Mechanical Tolerancing: The Results of an International Workshop. New York: ASME. 1988.

Andraesen, M. M., S. Kahler, T. Lund, with K. Swift. Design for Assembly. 2nd edition. United Kingdom: IFS Publications. 1988.

Baker, P. D., et al. A Roadmap to Manufacturing Excellence. Proceedings: 1987 IIE Integrated Systems Conference. Nashville, Tenn., November 1987.

Bebb, H. B., Quality design engineering: The missing link in U.S. competitiveness. Keynote address. National Science Foundation Engineering Design Conference. Amherst, Mass., June 1989. *A Xerox*

Corporation Vice-President describes the critical role of design quality in manufacturing competitiveness.

Berger, S., et al. Toward a new industrial America. Scientific American, Vol. 20, No. 9, June 1989, pp. 39-47.

Birmingham, W., A. Gupta, and D. P. Siewiorek. The micon system for computer design. IEEE Micro, October 1989, pp. 61-67, 1989.

Boothroyd, G., and P. Dewhurst. Design for Assembly—A Designer's Handbook. Technical Report. Department of Mechanical Engineering, University of Massachusetts. 1983.

Boothroyd, G., and P. Dewhurst. Product Design for Assembly Handbook. Wakefield, R.I.: Boothroyd Dewhurst, Inc. 1987. *Describes the essential concepts and practices of designing parts and products for economical assembly.*

Boothroyd, G., C. Poli, and L. March. Handbook of Feeding and Orienting Techniques for Small Parts. Technical Report. Mechanical Engineering Department, University of Massachusetts. 1978.

Box, G. E. P., and S. Bisgaard. The Scientific Context of Quality Improvement. Center for Quality and Productivity Improvement. University of Wisconsin-Madison. 1987. *A discussion of some Taguchi methods with proposals for simpler and more effective statistical procedures where appropriate.*

Box, G. E. P., J. S. Hunter, and W. G. Hunter. Statistics for Experimenters. New York: John Wiley & Sons. 1978.

Buzzel, R. D., and B. Gale. The PIMS Principles. New York: The Free Press. 1987.

Byrne, D. M., and S. Taguchi. The Taguchi approach to parameter design. Proceedings of the 1986 ASQC Quality Congress Transaction. 1986.

Chang, T. C., and R. A. Wysk. An Introduction to Automated Process Planning Systems. Englewood Cliffs, N.J.: Prentice-Hall. 1985. *This book describes coding and classification systems and variant and generative process planning. A description of the integration of CAD and CAM and the authors' Totally Integrated Process Planning System (TIPPS) are provided.*

Chase, K. W. Design issues in mechanical tolerance analysis. Manufacturing Review, Vol. 1, No. 1, March 1988, pp.50-59.

Chase, R. B., and D. A. Garvin. The Service Factory. Harvard Business Review, July-August 1985.

Clark, B. K., and T. Fujimoto. Overlapping problem solving in product development. Managing International Manufacturing. K. Ferdows, ed. Amsterdam: North-Holland. 1989.

Cutkosky, M. R., and J. M. Tenenbaum. CAD/CAM integration through concurrent process and product design. Pp. 1-10 in Intelligent and Integrated Manufacturing Analysis and Synthesis. New York: American

Society of Mechanical Engineers. 1987. *Describes research in simultaneous design of processes and products.*

Cutkosky, M. R., J. M. Tenenbaum, and D. Muller. Features in process-based design. ASME Computers in Engineering: Proceedings of the ASME International Computers in Engineering Conference and Exhibition. American Society of Mechanical Engineers, San Francisco, Calif., July 31 - August 3, 1988, pp. 557-562.

Dixon, J. R. Designing with features: Building manufacturing knowledge into more intelligent CAD systems. Proceedings of ASME Manufacturing International-88. American Society of Mechanical Engineers, Atlanta, Ga., April 17-20, 1988.

Dixon, J. R. On research methodology towards a scientific theory of engineering design. Artificial Intelligence for Engineering Design, Analysis, and Manufacturing (AIEDAM), Vol. 1, No. 3, June 1988. *An overview of scientific methods in the search for theoretical foundations for engineering design, prepared as a discussion base.*

Dixon, J. R., and M. R. Duffey. Quality is not accidental—It is designed. The New York Times, June 26, 1988, p. F2. *A concise statement of the critical role of design in competitiveness.*

Dixon, J. R., and M. R. Duffey. The neglect of engineering design. California Management Review, Vol. 32, No. 2, Winter 1990, pp. 9-23. *The coupling of design deficiencies to loss in competitiveness and a discussion of current and proposed responses.*

Dixon, J. R., J. J. Cunningham, and M. K. Simmons. Research in designing with features. IFIP WG 5.2 Workshop on Intelligent CAD Systems. D. Gossard, ed., Cambridge, Mass.: IFIP. 1987.

Dixon, J. R., M. R. Duffey, R. K. Irani, K. L. Meunier, and M. F. Orelup. A proposed taxonomy of mechanical design problems. Computers in Engineering: Proceedings of the ASME International Computers in Engineering Conference and Exhibition. American Society of Mechanical Engineers, San Francisco, Calif., July 31 - August 3, 1988, pp. 41-46.

Dornbusch, R. W., et al. The Case for Manufacturing in America's Future. Rochester, N.Y.: Eastman Kodak Company. 1987. *Discusses the importance of manufacturing vis-a-vis services in the national economy.*

Downey, W. G. Development Cost Estimating. Report of the Steering Group for the Ministry of Aviation, HMSO. 1969.

Elliot, J. G. Statistical Methods and Applications. Harper Woods, Mich.: Hamilton Printing Co. 1987. *A concise, straightforward tutorial on design of experiments, parameter design, tolerance design, and quality loss function.*

Finger, S., and J. R. Dixon. A Review of Research in Mechanical Engineering Design. Part I: Descriptive, Prescriptive, and Computer-Based Models of Design Processes; Part II: Representations, Analysis, and

Design for the Life Cycle. Research in Engineering Design, Part I, Vol. 1, No. 1, 1989, pp. 51-67, and Part II, Vol. 1, No. 2, 1989, pp. 121-137. *A comprehensive review covering more than 300 references.*

Fisher, R. A. Design of Experiments. New York: Hafner Publishing Co. 1951.

Foster, R. N. Innovation. New York: Summit Books. 1986.

Garvin, D. A. What does product quality really mean? Sloan Management Review, Fall 1984.

Gibbons, J. H. Advanced Materials by Design. Washington, D.C.: Congress of the United States. June 1988.

Gomory, R. E. Turning ideas into products. The Bridge, Spring 1988.

Gomory, R. E., and R. W. Schmitt. Science and product. Science, Vol. 240, May 27, 1988, pp. 1131-1204. *The authors argue that higher productivity stems at least in part from practice and learning and gradually improving products each time a new model is introduced. Faster product development cycles imply more accumulated improvements over time, opening a wider and wider productivity gap.*

Groover, M. P. Automation, production systems, and computer-integrated manufacturing: 1987. IIE Integrated Systems Conference, Nashville, Tenn., November 1987.

Hahn, G., and C. Morgan. Design experiments with your computer. Chemtech, November 1988.

Harry, M. J. The Nature of Six Sigma Quality. Government Electronics Group, Motorola, Inc. 1989.

Hauser, J., and D. Clausing. The house of quality. Harvard Business Review, May-June 1988, pp. 63-73. *Proposes a method for implementing the concepts of Quality Function Deployment.*

Hayes, R. H., S. C. Wheelwright, and K. B. Clark. Dynamic Manufacturing: Creating the Learning Organization. New York: The Free Press. 1988. *Describes the character, principles, structure, and operation of a continuously competitive manufacturing firm.*

Heginbotham, W. B. Programmable Assembly. United Kingdom: Springer-Verlag. 1984.

Inui, M., and F. Kimura. Representation and manipulation of design and manufacturing processes by data dependency. Intelligent CAD II: Proceedings of the IFIP TC 5/WG 5.2 Workshop on Intelligent CAD. H. Yoshikawa, T. Holden, eds. Amsterdam: Elsevier Science Publishers B.V. 1990.

Inui, M., H. Suzuki, F. Kimura, and T. Sata. Extending Process Planning Capabilities with Dynamic Manipulation of Product Models. Department of Precision Machinery Engineering, The University of Tokyo. 1987.

Jansson, D. G., and S. M. Smith. Design Fixation. Preprints of the 1989 NSF Engineering Design Research Conference, Amherst, Mass., June 1989.

Johnson, H. T. Activity-based information: Accounting for competitive excellence. Target, Spring 1989.

Kaplan, R. S. Management accounting for advanced technological environments. Science, Vol. 245, August 25, 1989, pp. 819-823. *Discusses the shortcomings of present financial cost accounting systems and presents the case for the activity-based costing (ABC) approach.*

Kaplan, R. S. Measures for Manufacturing Excellence. Boston, Mass.: Harvard Business School Press. 1990.

Kaplan, R. S. One cost system isn't enough. Harvard Business Review, May-June 1988.

Kerr, A. D., and R. B. Pipes. Why we need hands-on engineering education. Technology Review, October 1987, pp. 37-42.

Khalil, T. M., and B. A. Bayraktar. Challenges and Opportunities for Research in the Management of Technology. Workshop Report. NSF-University of Miami. February 1988.

Kimura, F., and H. Suziki. A CAD system for efficient product design based on design intent, Department of Precision Machinery Engineering, The University of Tokyo. Annals of the CIRP, Vol. 38, No. 1, 1989, pp. 149-152.

Leech, D. J., and B. T. Turner. Engineering Design for Profit. New York: John Wiley & Sons. 1985.

Libardi, E. C., J. R. Dixon, and M. K. Simmons. Computer environment for the design of mechanical assemblies: A research review. Engineering with Computers, Vol. 3, No. 3, 1988, pp. 121-136.

Maher, M. L. HI-RISE and beyond: Directions for expert systems in design. Computer-Aided Design, Vol. 17, 1985, pp. 420-427.

Meunier, K., and J. R. Dixon. Iterative respecification: A computational model for hierarchical mechanical system design. Computers in Engineering: Proceedings of the ASME International Computers in Engineering Conference and Exhibition. American Society of Mechanical Engineers, San Francisco, Calif., July 31 - August 3, 1988, pp. 25-32.

MIT Commission on Industrial Productivity. Made in America: Regaining the Productive Edge. Cambridge, Mass.: The MIT Press. 1989. *Results of an extensive study of the loss of competitiveness of American manufacturing industries with recommendations for changes.*

Miyakawa S., and T. Ohashi. The Hitachi assemblability evaluation method (AEM). Proceedings: First International Conference on Product Design for Assembly. Newport, R.I., April 1986. *Describes the Hitachi method of designing parts and products for economical assembly.*

Montgomery, D. C.   Design and Analysis of Experiments, 2nd edition. New York: John Wiley & Sons. 1984.

National Academy of Engineering.   Design and Analysis of Integrated Manufacturing Systems. Washington, D.C.: National Academy Press. 1988. *Chapter entitled "The Strategic Approach to Product Design" by D. Whitney et al. describes a concurrent design approach that focuses on the assembly process.*

National Academy of Engineering. Focus on the Future: A National Action Plan for Career-Long Education for Engineers. Washington, D.C.: National Academy Press. 1988. *The rationale for strengthening the career-long education of engineers and detailed action plans.*

National Academy of Engineering. The Technological Dimensions of International Competitiveness. Washington, D.C.: National Academy Press. 1988. *A broad examination of the issue of technology and competitiveness.*

National Research Council.   Computer Integration of Engineering Design and Production: A National Opportunity. Washington, D.C.: National Academy Press. 1984.

National Research Council. Engineering Education and Practice in the United States: Continuing Education of Engineers. Washington, D.C.: National Academy Press. 1985.

National Research Council. Engineering Education and Practice in the United States: Foundations of Our Techno-Economic Future. Washington, D.C.: National Academy Press. 1985.

National Research Council.   Management of Technology: The Hidden Competitive Advantage. Washington, D.C.: National Academy Press. 1987.

National Research Council. Systems Aspects of Cross-Disciplinary Engineering Research. Washington, D.C.: National Academy Press. 1986.

National Research Council.   Toward a New Era in U.S. Manufacturing: The Need for a National Vision. Washington, D.C.: National Academy Press. 1986.

National Science Foundation.   Report of the Workshop on Engineering Design. May 25-26, 1988. *The report of an industry-university-government panel that recommends action on improving engineering design, education, and research.*

Nevins J. L., and D. E. Whitney, eds.   Concurrent Design of Products and Processes. New York: McGraw-Hill. 1989. *A textbook on concurrent design, with a focus on design for assembly.*

Newsome, S. L. et al., eds.   Design Theory '88.   New York: Springer-Verlag. 1989.

Ostrofsky, B. Design, Planning, and Development Methodology. Englewood Cliffs, N.J.: Prentice-Hall, Inc. 1977.

Ouichi, W. G. The new joint R&D. Proceedings of the IEEE, September 1989, pp. 1318-1326.

Pahl, G., and W. Beitz. Engineering Design. London: The Design Council. New York: Springer-Verlag. 1984. *A landmark textbook of engineering design, published much earlier in Germany.*

Papalambros, P. Y., and D. J. Wilde. Principles of Optimal Design. Cambridge, England: Cambridge University Press. 1988.

Pennel, J. P., et al. The Role of Concurrent Engineering in Weapons System Acquisition. Institute for Defense Analysis Report R-338. 1988.

Pennel, J. P., and M. G. Slusarczuk. An Annotated Reading List for Concurrent Engineering. Institute for Defense Analysis. 1989. *Intended for both neophytes and experts in concurrent engineering; covers 120 works.*

Petroski, H. To Engineer Is Human: The Role of Failure in Successful Design. New York: St Martin's Press. 1982. *Engineering design from various points of view, a nontechnical essay that delights and informs readers with technical backgrounds.*

Phadke, M. S. Quality Engineering Using Robust Design. Englewood Cliffs, N.J.: Prentice-Hall. 1989.

Poli, C., J. Escudero, and R. Fernandez. How part design affects injection molding tool costs. Machine Design, November 24, 1988.

Polya, G. How To Solve It: A New Aspect of Mathematical Method. Princeton, N.J.: Princeton University Press. 1957.

Polya, G.. Patterns of Plausible Inference. Princeton, N.J.: Princeton University Press. 1954.

President's Commission on Industrial Competitiveness. Global Competition: The New Reality. 2 Vols. Washington, D.C.: U.S. Government Printing Office. 1985.

Rehfeldt, G. T. The return of competitiveness in American manufacturing companies— Lessons learned. SRI Meeting on The Strategic Management of Technology. San Francisco, Calif., January 26, 1988.

Requicha, A. A. G., and H. B. Voelcker. Solid Modeling: A Historical Summary and Contemporary Assessment. IEEE, Computer Graphics & Applications, March 1982, pp. 9-24.

Ryan, N. E. Tapping into Taguchi. Manufacturing Engineering, May 1987. *Describes the Taguchi approach to setting dimensions and tolerances for robust performance of products.*

Shah, J. J., and L. Pandit. Dezinev—An expert system for conceptual form design of structural parts. Computers in Engineering: Proceedings of the ASME International Computers in Engineering Conference and Exhibition. American Society of Mechanical Engineers, Chicago, Ill., 1986, pp. 17-24.

Shah, J. J., and P. R. Wilson. Analysis of knowledge abstraction, representation and interaction requirements for computer-aided engineering. Computers in Engineering: Proceedings of the ASME International Computers in Engineering Conference and Exhibition. American Society of Mechanical Engineers, San Francisco, Calif., July 31-August 3, 1988, pp. 17-24.

Sheehan, W. J., et al. The application of state-of-the-market CIM to GE's electrical distribution and control business. Electro 88 Conference Record. 1988.

Shephard, M. S., and M. A. Yerry. Approaching the automatic generation of finite element meshes. Computers in Mechanical Engineering, Vol. 1, No. 4, April 1983, pp. 49-56.

Squires, A. M. The Tender Ship: Governmental Management of Technological Change. Boston, Mass.: Birkhauser. 1986. Insights into the management structure and manager qualifications for technological enterprises.

Stalk, G. Time—The next source of competitive advantage. Harvard Business Review, July-August 1988.

Swift, K. G. Knowledge-Based Design for Manufacture. London: Kogan Page. 1987.

Taguchi, G. System of Experimental Design, Vol. 1 and Vol. 2. White Plains, N.Y.: UNIPUB/Kraus International Publications and Dearborn, Mich.: American Supplier Institute, Inc. 1987. Taguchi describes his methods for experimental design.

Trucks, H. E. Designing for Economical Production. Dearborn, Mich.: Society of Manufacturing Engineers. 1987.

Ullman, D. G. A Taxonomy of the Mechanical Design Process. Department of Mechanical Engineering, Oregon State University. 1988.

Ullman, D. G., and T. A. Dietterich. Mechanical design methodology. Computers in Engineering: Proceedings of the ASME International Computers in Engineering Conference and Exhibition. American Society of Mechanical Engineers, New York, 1988, pp. 173-180.

U.S. Congress, Office of Technology Assessment. Making Things Better: Competing in Manufacturing. OTA-ITE-443. Washington, D.C.: U.S. Government Printing Office. February 1990. An examination of largely nondesign issues in manufacturing and competitiveness.

Walker, E. A. Our engineering schools must share the blame for declining productivity. The Chronicle of Higher Education, December 2, 1987.

Whitney, D. E. Manufacturing by design. Harvard Business Review, July-August 1988.

Wilson, E. B. An Introduction to Scientific Research. New York: McGraw-Hill. 1952.

Wood, K. L., and E. K. Antonsson. Computations with Imprecise Parameters in Engineering Design: Background and Theory. Engineering Design Research Laboratory Report 88-01. California Institute of Technology. February 1988.

Zarefar, H., T. J. Lawley, and F. Etesami. PAGES: A parallel axis gear drive expert system. Computers in Engineering: Proceedings of the ASME International Computers in Engineering Conference and Exhibition. American Society of Mechanical Engineers, New York, 1986, pp. 145-149.

Notes

1. National Research Council, *Toward a New Era in U.S. Manufacturing* (Washington, D.C.: National Academy Press), 1986, p. 5.

2. S. Berger et al., "Toward a New Industrial America," *Scientific American*, Vol. 20, No. 9, June 1989, pp. 39-47.

3. President's Commission on Industrial Competitiveness, *Global Competition: The New Reality* (Washington, D.C.: U.S. Government Printing Office), 1985; U.S. Congress, Office of Technology Assessment, *Making Things Better: Competing in Manufacturing*, OTA-ITE-443 (Washington, D.C.: U.S. Government Printing Office), February 1990; and MIT Commission on Industrial Productivity, *Made in America: Regaining the Productive Edge*, (Cambridge, Mass.: The MIT Press), 1989.

4. Names other than product realization process are used for the process by which new and improved products are conceived, designed, produced, brought to market, and supported. The process includes determining customers' needs, translating those needs into engineering specifications, designing the product as well as its production and support processes, and operating those processes. Brief descriptions of this and other terms in this report appear in the Glossary, which begins on page 99.

5. The Profit Impact of Market Strategy data base, compiled by The Strategic Planning Institute of Cambridge, Mass., includes operating and quality data from approximately 3,000 business units in 450 companies for periods ranging from 2 to 10 years.

6. *The PIMS Principles* (New York: The Free Press), 1987.

7. J. R. Dixon and M. R. Duffey, "Quality Is Not Accidental—It Is Designed," *New York Times*, June 26, 1988.

8. Adapted from Chapter 1 of J. L. Nevins and D. E. Whitney, eds., *Concurrent Design of Products and Processes* (New York: McGraw-Hill),

1989. An earlier study giving similar results is reported in W. G. Downey, "Development Cost Estimating," *Report of the Steering Group for the Ministry of Aviation* (HMSO, 1969). Reference from D. J. Leech and B. T. Turner, *Engineering Design for Profit* (New York: John Wiley), 1985.

9. K. B. Clark and T. Fujimoto, "Overlapping Problem Solving in Product Development," K. Ferdows, ed., *Managing International Manufacturing* (Amsterdam: North-Holland), 1989.

10. In "Turning Ideas Into Products," *The Bridge*, Volume 18, No. 1, Spring 1988, pp. 11-14, R. E. Gomory, a senior vice-president of IBM, states that IBM's "most effective foreign competition has been characterized by tight ties between manufacturing and development, an emphasis on quality, the rapid introduction of incremental improvements . . . of preexisting product, and a tremendous effort *by those actually in the product cycle* to be educated on the relevant technologies, on the competition's products and on what is going on in the world."

11. The phrase "best engineering design practices" should be construed to mean the set of practices that is best for a particular company. Best practices will vary from firm to firm.

12. R. E. Gomory and R. W. Schmitt, *Science,* Vol. 240, May 27, 1988, pp. 1131-1204.

13. For example, R. S. Kaplan, "Management Accounting for Advanced Technological Environments," *Science,* Vol. 25, August 25, 1989, pp. 819-823.

14. U.S. Congress, Office of Technology Assessment, *Making Things Better: Competing in Manufacturing,* OTA-ITE-443 (Washington, D.C.: U.S. Government Printing Office), February 1990, Chapter 7.

15. See, for example, J. Hauser and D. Clausing, "The House of Quality," *Harvard Business Review*, May-June 1988, pp 63-73; R. B. Chase and D. A. Garvin, "The Service Factory," *Harvard Business Review*, July-August 1988, pp. 61-69; and G. Stalk, "Time-The Next Source of Competitive Advantage," *Harvard Business Review*, July-August 1988, pp. 41-53.

16. As noted earlier, various other names are also used for the product realization process.

17. See appendix for material provided by Polaroid and Hewlett-Packard describing their product realization processes.

18. The interplay of the various factors that enter into this phase of definition are particularly well described in papers by D. Garvin. See, for example, D. A. Garvin, "What Does Product Quality Really Mean?," *Sloan Management Review* 26, Fall 1984, p. 25.

19. The process of arriving at appropriate specifications is well described in J. Hauser and D. Clausing, "The House of Quality," *Harvard Business Review*, May-June 1988, pp. 63-73.

20. D. E. Whitney, "Manufacturing by Design," *Harvard Business Review*, July-August 1988, pp. 83-91.

21. Design practices used in this phase, which include designed experiments, Taguchi's robust design protocols, and specific programs such as Motorola's 6 sigma program, are described later in this chapter.

22. H. B. Bebb, "Quality Design Engineering: The Missing Link to U.S. Competitiveness," keynote address, National Science Foundation Engineering Design Conference, Amherst, Mass., June 1989.

23. Adapted from J. L. Nevins and D. E. Whitney, eds., *Concurrent Design of Products and Processes* (New York: McGraw-Hill), 1989, Chapter 8.

24. AT&T Bell Laboratories conducts research on product quality-cost models for semiconductor and printed wiring board design and fabrication processes. Research at Bell Labs yielded the Carter-Dishman theory that provides a guide to the economical application of VLSI, taking into account the many factors that enter into integrated circuit development and design.

25. J. Hauser and D. Clausing, "The House of Quality," *Harvard Business Review*, May-June 1988, pp 63-73.

26. R. N. Foster, *Innovation* (New York: Summit Books), 1986.

27. See, for example, Manufacturing Studies Board, *Toward a New Era in Manufacturing* (Washington, D.C.: National Academy Press), 1986; R. S. Kaplan, *Measures for Manufacturing Excellence*, (Boston: Harvard Business School Press), 1990, and "Management Accounting for Advanced Technological Environments," *Science*, August 25, 1989, p. 819 ff.

28. Sometimes called "quadratic-loss-function," a somewhat inappropriate name since not all qlfs are quadratic and the utility is vastly broader than that for the quadratic case.

29. The conflicts that can arise because of differing quality definitions among the various functional organizations in a firm are discussed in D. A. Garvin, "What Does Product Quality Really Mean?," *Sloan Management Review* 26, Fall 1984, p. 25.

30. M. J. Harry, "The Nature of Six Sigma Quality," Government Electronics Group, Motorola, Inc.

31. Recent references on current DFM and DFA techniques are K.G. Swift, *Knowledge-Based Design for Manufacture* (London: Kogan Page), 1987; M. M. Andraesen, S. Kahler, T. Lund, with K. Swift, *Design for Assembly*, 2nd edition, (United Kingdom: IFS Publications), 1988.

32. S. Miyawaka and T. Ohashi, "The Hitachi Assemblability Evaluation Method" (now the Hitachi Producibility Method), *Proceedings 1st International Conference on Product Design for Assembly*, Newport, R.I., April 1986; G. Boothroyd and P. Dewhurst, *Product Design for Assembly Handbook* (Wakefield, R.I.: Boothroyd Dewhurst, Inc.), 1987.

33. D. E. Whitney, "Manufacturing by Design," Harvard Business Review, July-August 1988, pp. 83-91; J. L. Nevins and D. E. Whitney, eds.,

Concurrent Design of Products and Processes (New York: McGraw-Hill), 1989.

34. W. J. Sheehan et al., "The Application of State-of-the-Market CIM to GE's Electrical Distribution and Control Business," *Electro 88 Conference Record*, 1988.

35. G. T. Rehfeldt, "The Return of Competitiveness in American Manufacturing Companies—Lessons Learned," SRI Meeting on the Strategic Management of Technology, San Francisco, Calif., January 26, 1988.

36. There are several statements of the Principle of Robust Design. M. S. Phadke states it as, "Minimize the *effect* of the cause of variation without controlling the cause itself." J. G. Elliott says, "Americans remove the cause of the effect. Japanese remove the effect of the cause."

37. As described in D. M. Byrne and S. Taguchi, "The Taguchi Approach to Parameter Design", *Proceedings of the 1986 ASQC Quality Congress Transaction*, 1986.

38. A good exposition and examples of the technique are provided in M. S. Phadke, *Quality Engineering Using Robust Design* (Englewood Cliffs, N.J.: Prentice-Hall), 1989.

39. There are many books and references on SPICE in its various versions, such as P. Tuinenga, *SPICE: A Guide to Circuit Simulation and Analysis, P-SPICE* (New York: Prentice-Hall), 1988. A good survey paper containing a historic account of the development of SPICE in its several forms is contained in a paper by A. Vladimierescu, *Proceedings of the Bipolar Circuits and Technology Meeting*, September 1990.

40. G. Hahn and C. Morgan, "Design Experiments with Your Computer," *Chemtech*, November 1988. The American Supplier Institute, Dearborn, Michigan, provides PC-based software that helps in the design and guides the execution and analysis of design experiments using Taguchi's techniques. Texas Instruments has announced a PC-based expert system that will make it possible for the user to conduct experiments of this type with no additional training.

41. R. A. Fisher, *Design of Experiments* (New York: Hafner Publishing Co.), 1951.

42. See G. E. P. Box, J. S. Hunter, and W. G. Hunter, *Statistics for Experimenters* (New York: John Wiley & Sons), 1978; and D. C. Montgomery, *Design and Analysis of Experiments*, 2nd ed. (New York: John Wiley & Sons), 1984.

43. M. S. Phadke, *Quality Engineering Using Robust Design*, describes these techniques and provides examples; J. G. Elliott, *Statistical Methods and Applications*, is a Taguchi "cookbook" that describes how to apply this method using examples from the automobile industry; Taguchi's contributions and the relationship between Taguchi's methods and traditional design of experiments are described clearly in G. E. P. Box and S. Bisgaard, *The*

Scientific Context of Quality Improvement (Madison, Wisc.: Center for Quality and Productivity Improvement, University of Wisconsin), 1987.

44. Personal communication from Dr. J. Spira, president, Lutron Electronics Co. Inc., Coopersburg, Pa.

45. M. Patterson, Director of Corporate Engineering at Hewlett-Packard, believes that he can identify people with innate abilities for design by "their patterns of analogic thought."

46. E. B. Wilson, *An Introduction to Scientific Research* (New York: McGraw-Hill), 1952.

47. G. Polya, *How To Solve It: A New Aspect of Mathematical Method* (Princeton, N.J.: Princeton University Press), 1957; and *Patterns of Plausible Inference* (Princeton, N.J.: Princeton University Press), 1954.

48. H. Petroski, *To Engineer Is Human—The Role of Failure in Successful Design* (New York: St. Martin's Press), 1982.

49. National Research Council, Panel on Continuing Education of the Committee on the Education and Utilization of the Engineer, *Engineering Education and Practice in the United States: Continuing Education of Engineers* (Washington, D.C.: National Academy Press), 1985.

50. There is direct evidence of this effect in universities that have hired recent graduates whose research was supported by the NSF Design Theory and Methodology Program.

51. National Research Council, *Engineering Education and Practice in the United States, Foundations of Our Techno-Economic Future* (Washington, D.C.: National Academy Press), 1985, pp. 61-63.

52. Among others, National Research Council, *Engineering Education and Practice in the United States, Foundations of Our Techno-Economic Future*, 1985; ABET, "Engineering Education Answers the Challenge of the Future," *Proceedings of the National Congress on Engineering Education*, 1986; ASEE, *A National Action Agenda for Engineering Education*, 1987; National Science Foundation, *Report of the Workshop on Engineering Design*, May 25-26, 1988; A. D. Kerr and R. B. Pipes, "Why We Need Hands-on Engineering Education," *Technology Review*, October 1987.

53. The latter ability sometimes stands out because many older engineers are not proficient with computers.

54. Accreditation Board for Engineering and Technology, Inc., *Annual Report*, 1989.

55. National Academy of Engineering, *Focus on the Future: A National Action Plan for Career-Long Education for Engineers*, Report of the Committee on Career-Long Education for Engineers (Washington, D.C.: National Academy Press), 1988; National Research Council, *Engineering Education and Practice in the United States: Continuing Education of Engineers* (Washington, D.C.: National Academy Press), 1985.

56. In Germany, one must have several years industrial experience before being admitted to an engineering faculty.

57. The best known of these is G. Pahl and W. Beitz, *Engineering Design*, The Design Council (London: Springer-Verlag), 1984.

58. IGES stands for International Graphics Exchange Standard; PDES stands for Product Data Exchange Specification.

59. This was one of the recommendations of the National Science Foundation's 1988 Workshop on Engineering Design.

60. For comparison, the practice of medicine today is based not only on heuristics, but also upon a great deal of basic research in the fields of physiology, biology, physics, and pharmacology. As a result, medical practice today is far superior to late nineteenth century practice, which was based on an evolving collection of unscientific heuristics (some of which worked, or appeared to work sometimes). Closer to current engineering practice, the basic research effort in materials over the past three decades has produced a number of general principles and resulted in striking advances in practical new materials, new industries, and a cadre of highly productive materials engineers and scientists.

61. D. G. Jansson and S. M. Smith, "Design Fixation," *Preprints of the 1989 NSF Engineering Design Research Conference*, Amherst, Mass., June 1989.

62. Further categorization based on size, complexity, technical level, and other factors is also possible.

63. One methodology well known and well established for certain parametric design problems is optimization. Another body of knowledge useful in parametric design is statistical design of experiments (Taguchi makes application of these methods). Other knowledge-based approaches to parametric design of components, often implemented in computer programs, have also been developed. These are further referenced and discussed in Chapter 2.

64. For example, because there are formal means such as optimization and statistics, it can be said that there almost exists at present a theory of parametric design of components. This cannot yet be said for the other design problem categories identified above.

65. E. C. Libardi, J. R. Dixon, and M. K. Simmons, "Computer Environment for the Design of Mechanical Assemblies: A Research Review," *Engineering with Computers*, Vol. 3, No. 3, 1988, pp. 121-136.

66. J. J. Shah and P. R. Wilson, "Analysis of Knowledge Abstraction, Representation and Interaction Requirements for Computer-Aided Engineering," *Computers in Engineering: Proceedings of the ASME International Computers in Engineering Conference and Exhibition* (San Francisco, Calif.: American Society of Mechanical Engineers), July 31-August 3, 1988, pp. 17-24.

67. A. A. G. Requicha and H. B. Voelcker, "Solid Modeling: A Historical Summary and Contemporary Assessment," *IEEE, Computer Graphics & Applications*, March 1982, pp. 9-24.

68. J. R. Dixon, J. J. Cunningham, and M. K. Simmons, "Research in Designing with Features," *IFIP WG 5.2 Workshop on Intelligent CAD Systems*, D. Gossard, ed. (Cambridge, Mass.: IFIP), 1987.

69. M. R. Cutkosky and J. M. Tenenbaum, "CAD/CAM Integration Through Concurrent Process and Product Design," *Intelligent and Integrated Manufacturing Analysis and Synthesis* (New York: American Society of Mechanical Engineers), 1987, pp. 1-10; J. R. Dixon, "Designing with Features: Building Manufacturing Knowledge into More Intelligent CAD Systems," *Proceedings of ASME Manufacturing International-88* (Atlanta, Ga.: American Society of Mechanical Engineers), April 17-20, 1988.

70. D. G. Ullman and T. A. Dietterich, "Mechanical Design Methodology," *Computers in Engineering: Proceedings of the ASME International Computers in Engineering Conference and Exhibition* (New York: American Society of Mechanical Engineers), 1988, pp. 173-180.

71. J. R. Dixon, M. R. Duffey, R. K. Irani, K. L. Meunier, and M. F. Orelup, "A Proposed Taxonomy of Mechanical Design Problems," *Computers in Engineering: Proceedings of the ASME International Computers in Engineering Conference and Exhibition* (San Francisco, Calif.: American Society of Mechanical Engineers), July 31-August 3, 1988, pp. 41-46; D. G. Ullman, "A Taxonomy of the Mechanical Design Process," personal communication, Oregon State University, 1981.

72. M. L. Maher, "HI-RISE and Beyond: Directions for Expert Systems in Design," *Computer-Aided Design*, Vol. 17, 1985, pp. 420-427; J. J. Shah and L. Pandit, "Dezinev—An Expert System for Conceptual Form Design of Structural Parts," *Computers in Engineering: Proceedings of the ASME International Computers in Engineering Conference and Exhibition* (Chicago, Ill.: American Society of Mechanical Engineers), 1986, pp. 17-24; H. Zarefar, T. J. Lawley, and F. Etesami, "PAGES: A Parallel Axis Gear Drive Expert System," *Computers in Engineering: Proceedings of the ASME International Computers in Engineering Conference and Exhibition* (New York: American Society of Mechanical Engineers), 1986, pp. 145-149.

73. P. Y. Papalambros and D. J. Wilde, *Principles of Optimal Design* (Cambridge, England: Cambridge University Press), 1988.

74. K. Meunier and J. R. Dixon, "Iterative Respecification: A Computational Model for Hierarchical Mechanical System Design," *Computers in Engineering: Proceedings of the ASME International Computers in Engineering Conference and Exhibition* (San Francisco, Calif.: American Society of Mechanical Engineers), July 31-August 3, 1988, pp. 25-32.

75. G. Taguchi, *System of Experimental Design*, Vol. 1 and Vol. 2 (White Plains, N.Y.: UNIPUB/Kraus International Publications, and Dearborn, Mich.: American Supplier Institute, Inc.), 1987.

76. K. W. Chase, "Design Issues in Mechanical Tolerance Analysis," *Manufacturing Review*, Vol. 1, No. 1, March 1988, pp. 50-59.

77. M. S. Shephard and M. A. Yerry, "Approaching the Automatic Generation of Finite Element Meshes," *Computers in Mechanical Engineering*, April 1983, pp. 49-56; A. M. Agogino and A. S. Almgren, "Symbolic Computation in Computer-Aided Optimal Design," *Expert Systems in Computer-Aided Design*, J. S. Gero, ed. (Amsterdam: North-Holland), 1987, pp. 267-284; K. L. Wood and E. K. Antonsson, "Computations with Imprecise Parameters in Engineering Design: Background and Theory," Engineering Design Research Laboratory Report 88-01, California Institute of Technology, February 1988.

78. W. Birmingham, A. Gupta, and D. P. Siewiorek, "The Micon System for Computer Design," *IEEE Micro*, October 1989, pp 61-67.

79. For good descriptions of variant process systems, see M. Inui and F. Kimura, "Representation and Manipulation of Design and Manufacturing Processes by Data Dependency," *Intelligent CAD II: Proceedings of the IFIP TC 5/WG 5.2 Workshop on Intelligent CAD*. H. Yoshikawa and T. Holden, eds. (Amsterdam: Elsevier Science Publishers B.V.), 1990; F. Kimura and H. Suziki, "A CAD System for Efficient Product Design Based on Design Intent," Department of Precision Machinery Engineering, The University of Tokyo, *Annals of the CIRP*, Vol. 38, No. 1, 1989, pp. 149-152; M. Inui, H. Suzuki, F. Kimura, and T. Sata, "Extending Process Planning Capabilities with Dynamic Manipulation of Product Models," Department of Precision Machinery Engineering, The University of Tokyo, 1987.

80. M. Cutkosky, J. Tenenbaum, and D. Muller, "Features in Process-Based Design," *Computers in Engineering: Proceedings of the ASME International Computers in Engineering Conference and Exhibition* (San Francisco, Calif.: American Society of Mechanical Engineers), July 31-August 3, 1988, pp. 557-562.

81. D. E. Whitney, J. L. Nevins, T. L. DeFazio, R. E. Gustavson, R. W. Metzinger, J. M. Rourke, and D. S. Seltzer, "The Strategic Approach to Product Design," *Design and Analysis of Integrated Manufacturing Systems* (Washington, D.C.: National Academy Press), 1988, pp. 200-223.

82. G. Boothroyd, C. Poli, and L. March, "Handbook of Feeding and Orienting Techniques for Small Parts," Technical Report, Mechanical Engineering Department, University of Massachusetts, 1978; G. Boothroyd and P. Dewhurst, "Design for Assembly—A Designer's Handbook," Technical Report, Department of Mechanical Engineering, University of Massachusetts, 1983.

83. C. Poli, J. Escudero, and R. Fernandez, "How Part Design Affects Injection Molding Tool Costs," *Machine Design*, November 24, 1988.

84. D. Clausing and J. R. Hauser, "The House of Quality," *Harvard Business Review*, May-June 1988, pp. 63-73.

85. R. S. Kaplan, "One Cost System Isn't Enough," *Harvard Business Review*, May-June 1988; and R. S. Kaplan, "Managerial Accounting for Advanced

Technological Environments," *Science*, Vol. 245, August 25, 1989, pp. 819-833.

86. D. Clausing and J. R. Hauser, "The House of Quality," *Harvard Business Review*, May-June 1988, pp. 63-73.

87. T. J. Allen, *Managing the Flow of Technology* (Cambridge, Mass.: MIT Press), 1977.

88. Two recent examples are the design for assembly work of Professor Boothroyd and the solid modeling foundations laid by Professors Voelcker and Requicha. In the former, university research identified the crucial geometric abstractions needed to predict handling and assembly costs, and these were then translated into specific design support tools. In the latter, the formal mathematical foundations for representations of solid objects were developed, and these were used as the basis not only for early versions of PADL, a pioneering solid modeler, but also for much of the solid modeling capability that has evolved since.

89. For example, a number of recent graduates whose research was supported by the NSF Design Theory and Methodology Program have gone to work in design-related positions with forefront firms and educational institutions.

90. This is the only established route to date. The category of design-oriented companies includes ComputerVision, Parametric Engineering, ICAD, Intellicorp, and Carnegie Group.

91. In other major competitive nations (e.g., Japan and Germany), mechanisms for performing and sharing such applied research are well established. The Fraunhofer Institutes in Germany are examples.

92. U.S. Congress, Office of Technology Assessment, *Making Things Better: Competing in Manufacturing*, OTA-ITE-443 (Washington, D.C.: U.S. Government Printing Office), February 1990, pp 73-74.

93. Some of these are discussed in U.S. Congress, Office of Technology Assessment, *Making Things Better: Competing in Manufacturing*, OTA-ITE-443 (Washington, D.C.: U.S. Government Printing Office), February 1990, pp 202-211.

Index